ZERO LATENCY LEADERSHIP

BEENA AMMANATH

ZERO LATENCY LEADERSHIP

Driving Equity, Trust, and Sustainability with Emerging Tech

Forbes | Books

Published by Forbes Books, Charleston, South Carolina.
Member of Advantage Media.

Printed in the United States of America.

10 9 8 7 6 5 4 3 2 1

ISBN: 979-8-88750-058-4 (Hardcover)
ISBN: 979-8-88750-059-1 (eBook)

LCCN: 2023901862

Cover design by Analisa Smith.
Layout design by David Taylor.

To my parents, Kamalam and Kumar,
who gave me the wings to fly

CONTENTS

PART THREE

FOREWORD

It's no exaggeration that technology has radically changed the world around us in the last few decades. Traditional business models have been disrupted and entirely new industries have been created—all fueled by tech. Yet it seems that the changes that we all have experienced so far will likely be dwarfed by the changes yet to come, as technology innovation continues to accelerate.

Organizations have also seen the unintended consequences new technologies can have on society. Businesses globally are still trying to catch up on solving for ethics and trust in artificial intelligence. But we have a unique opportunity in front of us, for some of the other newer emerging technologies like the metaverse, blockchain, quantum and digital reality—to proactively identify and prioritize for trust before it scales across all our enterprises. This is not only important from a business risk perspective but also for all society's collective future.

Businesses need leadership to keep ethics and trust top of mind as new emerging technologies enter their organizations. A leadership that melds the conversation about competitive advantage with trust, inclusivity, and sustainability. A leadership that leverages rapidly emerging technologies for business advancement while also considering the impact to our communities and the world. A leadership that is about the knowledge, patience, and attitude necessary to make mindful leadership decisions. We need leadership that is fluent to navigate the complex technology landscape, consider the advantages and disadvantages of each technology by itself and when these technologies blend together.

This book provides readers with an opportunity to think forward while rooted to the present and aware of both the great accomplishments and the mistakes of our collective past with technology. We have the enormous privilege of living in this time of massive innovation and access to new technologies. We also have the great responsibility of choosing wisely while not missing the opportunity as leaders to build stronger and more successful organizations. By revisiting the past, establishing a decision-making context, and building technical literacy, lets welcome a new horizon of leadership—one which is trustworthy, sustainable, inclusive, and benefits both businesses and society.

KWASI MITCHELL
Chief Purpose Officer
Deloitte US

PART ONE

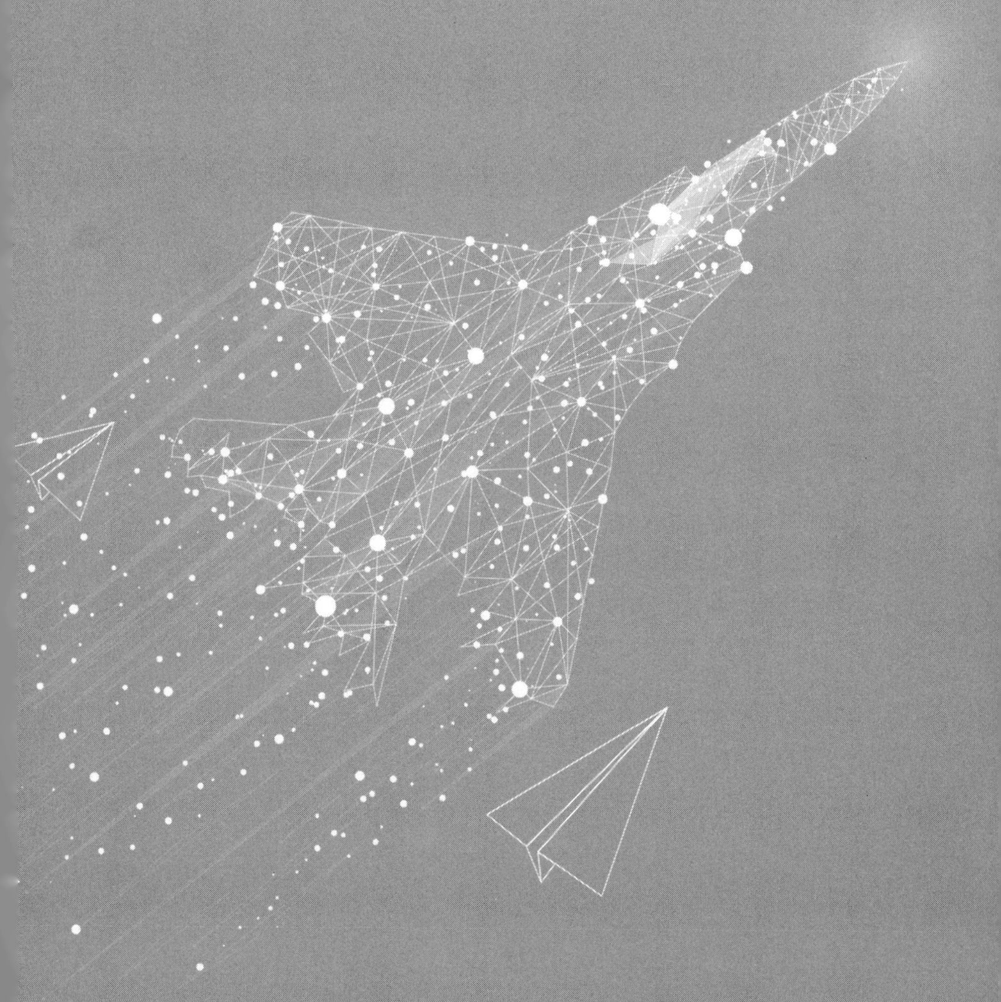

INTRODUCTION

We live in a time when every decision we make as leaders is simultaneously under a microscope, on a virtual billboard, and recorded for posterity in a matter of seconds for all the world to see. Yet we must make decisions faster than ever because the technology and circumstances of business demand it. Under these modern pressures, a corresponding tool set is necessary to navigate these challenges.

Zero Latency Leadership allows leaders to align tech resources with trust, equity, and sustainability to shape reliable competitive advantages for their business. This book will help you build a personal leadership framework for making complex decisions at great speed and with purpose. To move fluidly in the moment, we must revisit the lessons of the past and frame our decisions and practices in a light that builds trust and creates space for learning and growth. By integrating wisdom, knowledge, and opportunity, we build a better future through informed and mindful leadership.

This book is an opportunity to think forward while being rooted to the present, aware of both the great accomplishments and tragic mistakes of our collective past. We have the great privilege of living in this time of massive innovation and access to new technologies.

We also have the great responsibility of choosing wisely while not missing the opportunity as leaders to build stronger and more successful organizations.

Regardless of your domain, technology is the driving force of modern business infrastructure and must be front of mind in all strategy and execution. Our dependency on technology has never been greater than it is right now, and this dependency grows daily. The global pandemic was a crucible that clarified just how integral technology has become to every kind of business.

Technology is the driving force of modern business infrastructure and must be front of mind in all strategy and execution. Our dependency on technology has never been greater than it is right now, and this dependency grows daily.

Economies and businesses that weathered and are still weathering the economic upheavals of the 2020s survived on lifeboats of technology. The invisible technology that underpins our society has become center stage, and it's not going to fade into the background.

Films that could not be released to theaters were streamed to homes. Employees who could not report to the office continued to work while they sheltered in place. E-commerce boomed as people were unable to shop in person. Children graduated from high school and matriculated to college from kitchen-counter campuses. The pandemic was a flashpoint, but the lessons it taught us are lasting. The great resignation and economic recessions that followed the disruptions of the healthcare crisis only strengthened the markets for technology and innovative approaches to problems both new and old.

Zero Latency Leaders assess and adopt the technology they need while navigating the rough waters of a distrustful world. Businesses that cannot count on fundamental values while embracing the necessary flexibility to adapt to the new models of work and commerce will fail to adapt.

Foundations: Trust, Equity, and Sustainability

The ethical and values-based concepts of trust, equity, and sustainability are the foundations of modern credibility. Without credibility, our leadership decisions are suspect until proven solid, and our value propositions are shaky at best. We need our credibility to steer our organizations through the rough waters of contemporary times. The downside of an open society paired with infinite communication channels is the proliferation of misinformation. Misinformation erodes trust and polarizes people. Distrust and mistrust are not just disruptive; they endanger business performance with customers and employees alike. Having the trust of stakeholders is essential to move forward and to have the space to make difficult decisions.

Lasting trust is built on a foundation of honesty, integrity, and thoughtfulness. Equity, which is built on inclusion and fair practices, is at the heart of credibility. Equity and diversity are not simply buzzwords; they are essential components to building a lasting business presence. Exclusion is an artifact of a small-minded approach to business. Modern technology has made our global connectedness the greatest opportunity for business, and we dismiss that at our peril. The more voices we engage, the better our products and

Equity and diversity are not simply buzzwords; they are essential components to building a lasting business presence. Exclusion is an artifact of a small-minded approach to business.

customer loyalty will become. Likewise, employee engagement and retention are only possible with an equitable approach. Our technology decisions as leaders must take equity into account to ensure they are credible and successful.

Sustainability is another term that needs more than lip service. Reducing our carbon footprint and ecological impact is no longer a "nice to have" element of business. We must demonstrate that sustainability, ecological, financial, and social impact inform all our decisions. Pleading ignorance in the information age is insincere at best and corrosive to credibility and trust.

Stakeholders trust leaders who understand the risks and rewards of their decisions. Zero Latency Leaders think through the variety of impacts their choices may have and take ownership for their decisions. Their choices are deliberate, and they accept the risks as well as the rewards, not just because of the optics of the moment but also because of the impact that will be felt by those affected now and in the future. Zero Latency Leadership provides examples of *why* deliberation and appraisal of consequence is so crucial in making decisions about technology and *how* abandoning the guiding principles of equity, sustainability, and credibility endangers not only reputations and business but also all of society. Likewise, these principles will help build trust and draw support from employees, shareholders, and customers alike.

Getting to Zero

You would not be where you are today without a strong foundation of leadership, and those skills and abilities will serve you well as you level up your leadership style to meet the challenges and opportunities of emerging technologies. Specifically, your agility or ability to adapt as a leader will be critical to approaching the ever-changing nature of technology. That core agility, blended with an innovative spirit, critical thinking, and strong decision-making skills, will prepare you to address the difficult choices ahead while having the resilience to learn from mistakes.

None of your agility skills will matter much, however, if you do not also hone your interpersonal skills. Negotiation and conflict management are essential, as some people will be fearful of rapid change or the temporary failure that accompanies experimentation. To keep employees and stakeholders motivated through the necessary changes, you must lead with empathy. As a leader, when you build relationships in which honesty is the foundation of your mutual success, trust will follow. Zero Latency Leadership enhances your already existing leadership skills that got you to where you are today to succeed tomorrow.

This book is designed to set you on your path to Zero Latency Leadership. You will build your technical literacy and learn about emerging technological innovations and what they mean for business leadership.

Part one establishes a context, both historical and contemporary, for the tools, lessons, and resources available to position trustworthy leadership within technology.

Part two introduces a model of applied leadership through a fictitious company, BA Biotech (BAB). After each applied leadership example, we will build on your technical literacy by explaining some of the leading and emerging technologies you will encounter as an executive. Each chapter in part two will establish the risks and rewards of embracing that technology, ending with some thought-provoking questions about how that technology may fit into your business model and your professional success.

Part three will introduce more technologies to consider for the future state to round out a primer for technical literacy. Then we will focus on the characteristics of a Zero Latency Leader and the next steps on your leadership journey.

At its heart this book is about preparing you to face the future empowered by a healthy appraisal of risk and reward and unhampered by fears. We all walk before we run, and each movement strengthens our abilities. So inspired, let's take our first step together on the path to the Zero Latency Leader.

Complexity should not frighten us; impatience is always the greater danger.

—DR. SARA TERHEGGEN

HISTORY OF CHANGE

While this book is primarily focused on the technology landscape, history offers us a treasure trove of examples to help us frame our perspective as we move forward with the next evolution of technology. For thousands of years, human beings could only imagine the technologies we use daily in the 2020s. Yet we got to where we are from the skills that were commonplace throughout human history: the use of fire and basic tools, the power of observation, and verbal communication. On the way to present day, our ancestors developed the written language, the printing press, mass production, the steam engine, the combustion engine, airplanes, submarines, space travel, DNA analysis, organ transplant, life support, and the internet.

During the twentieth century, human beings witnessed the rapid creation and obsolescence of technologies that created crazes, booms, and busts. From horse-drawn carriages to the first electric car, no previous century generated more mobilization or changes in how human beings lived daily life. The century began with most people on the planet living on farms without electricity or indoor plumbing, and even telephones were luxuries rarely seen by any but the wealthy in larger towns and cities.

In 1900 people washed their laundry on rocks or on washboards, bathed in water pumped or drawn from wells, and used outdoor latrines in most places on earth. Children born in 1900, still living in 2000, witnessed two World Wars, the first of which was fought with horses, and the second ended with the atomic bomb. This same generation saw the inventions of the wireless radio, the talking picture, the television, and the internet. Though now they are all gone, those people who shared this earth with us just over two decades ago would have heard about the flight from Kitty Hawk as small children in 1903 from their family and friends who had read about in the newspaper days or weeks after it happened. The same small children of 1903 then watched Neil Armstrong walk on the moon on their televisions in 1969. The world changed substantially, and humans adapted remarkably quickly to those changes. In many ways the world changed for the better, with faster communication, better agricultural practices, longer life spans, lower infant mortality, and more comfortable living conditions. However, that change did not come without consequences, many of which we are still contending with today.

Here we are, nearly one-quarter of the way through the twenty-first century, marveling at how much our own lives have changed since leaving the twentieth century. Yet when we worry about technology today, we can take some comfort that we are hardly the first generation to see so much change. Humanity has been changing and evolving since the beginning of recorded time.

Mobilization and change are at the very core of human history, and it is human nature to record that history. The people who first sailed to the Hawaiian Islands as well as the Natives of the Americas who first crossed the Bering Strait, and all human migrations since, have bred some forms of innovation. We can learn from ancient wisdom that we are but a segment on a long line, neither at the beginning nor at the end of evolution.

Diffusion of Technology

Since 1983, the measure of technology adoption has used the Rogers scale,[1] which includes innovators, early adopters, the early majority, the late majority, and laggards. These labels are used to identify the reactive behaviors of individuals when a new technology becomes available to them. The Rogers scale, and the tools used to measure for it, has been applied to everything from the use of automatic teller machines (ATMs) to Wi-Fi adoption in homes. It has also been used to investigate the historical adoption of the telephone in different regions and the adoption of the printing press throughout central Europe. It is an effective tool for historical and contemporary insights. Here is a contemporary model to help illustrate the concept.

1 Jeffrey James, "The Diffusion of IT in the Historical Context of Innovations from Developed Countries," *Social Indicators Research*, 2013, https://www.ncbi.nlm.nih.gov/pmc/articles/PMC3560945/.

Consider the person who waits in line overnight to buy the very first iPhone and every iteration after it the moment it comes to market. Those people sleeping in line in front of the Apple Stores are innovators. The people who wait to buy a fully debugged next edition of the original iPhone and only upgrade to the next but most reliable iteration are early adopters. The multitude of teenagers so eager to get their hands their first iPhones are the early majority, who see the status over the utility of adopting the technology.

Then there are the bankers who finally give up their Blackberries to have the same phones as the rest of their families; they make up the late majority. Finally, we have the now disenfranchised family member, who is now entirely out of the loop with their landline. This laggard may inherit a used iPhone or get a bare-bones smartphone just so they can text and know when plans have changed.

Now consider the first people to get actual landline phones after Alexander Graham Bell brought them to market. Bell's first successful phone call happened in 1876, and within three years, nearly forty-nine thousand telephones were installed. That number increased more than tenfold by the turn of the twentieth century, and it continued to grow, but over the course of more than fifty years. Ultimately, only laggards, either because of lack of access or a dislike for technology, did not have telephones in their homes or businesses.[2]

We can look at the historical adoption of technologies to better understand how new technologies will affect the population.

Technology spreads at different rates based on the cost, complexity, and usefulness and utility of the concepts and tools involved. Pre-

2 Elon University, "1870s–1940s: Telephone," *Imaging the Internet,* January 17, 2023, https://www.elon.edu/u/imagining/time-capsule/150-years/back-1870-1940/#:~:text=Bell%20began%20his%20research%20in,49%2C000%20telephones%20were%20in%20use.

internet personal computers didn't take off until the user interfaces (UIs) made them easier for the average person to operate, and mass production of silicon chips made them more affordable.

We can look at the historical adoption of technologies to better understand how new technologies will affect the population. From the innovators who will grasp the newest technologies, hack them, exploit them, and discover new ways to use them to the laggards who will wait it out until they have no choice but to join the club, every wave of technology will wash over us at some point.

The studies of technology diffusions are as endless at the types of technologies that can be measured. It's not the purpose of this book to perform a literature review of diffusion of technology, but it is a crucial concept to keep in mind as we look at the past, present, and future of human behavior as it intersects the technology on our horizon. We should be reassured by the fact that the questions we ask about the newest types of technology are the same questions that were asked about the technologies we use every day and the ones that are no longer used because they are now obsolete.

Everything Old Is New Again

Privacy was an early and lasting concern about the original telephone. In fact, most of the regulation on the current internet had its genesis with the telephone. Regulations about unsolicited calls, harassment, and scams have been concerns since the early twentieth century. So it should come as no surprise that every technology hence has shared the same concerns. We can learn from over 150 years of telephony history to inform our approach to the future of technology.

It doesn't stop with the telephone. The automobile also has many lessons to teach us about unintended consequences and how regulation emerged. Before Henry Ford built his first factory, the automobile had existed for decades in Europe. Early drivers were the innovators at the front edge of the diffusion curve, and like most innovators, they had money to spend on experimental technology. It was the convergence of one technology (the operational automobile) with another innovation (the assembly line) that led to the proliferation of automobiles across the globe.

The volume of car ownership literally changed the shape of the world in so many ways. Roman roads in Europe still existed long before cars were popular. Those roads determined the size of the chassis of wagons, carriages, and eventually, the horseless carriage. Those same dimensions in turn informed the size of the modern automobile and the modern road.

Paved parking, interstate highways, and toll transponders are all innovations that exist only because of their symbiotic relationship to the mass production of the automobile. Likewise, semitruck weigh stations, alcohol limits, and safety legislation that all drivers must learn about before they ever get behind the wheel are evolutions of driving culture that followed the mass production of the automobile.

> **The social, economic, and environmental impacts of any technology come in waves, and we would be wise to remember that.**

While mass production is not as major a contributor to the proliferation of air travel as it is to automotive travel, parallels between the cultural impacts of both air and automotive travel do exist. Air travel has led to many similar and subsequent inventions and innovations. Airplanes didn't have much consumer appeal until they could carry multiple passengers safely over significant distances. And they still would have little value without airports.

A rush to build airports in major metropolitan areas had unintended consequences on property values as planes grew larger and louder with the increased amount of air traffic. Likewise, air traffic patterns impact wildlife through noise, air, and ground pollution as well as disruption from airport construction and traffic to and from those locations.

As train travel and the automobile both led to tourism in local areas, those venues were quickly supplanted by air travel taking tourists farther afield. Affordable air travel made remote tourist destinations like the Hawaiian and Caribbean islands far more accessible for people across the globe. This in turn led to the decline of scenic spots like the Catskills and Great Lakes that had been local favorites earlier in the twentieth century. This was true for many European seaside resorts and beauty spots as well. The social, economic, and environmental impacts of any technology come in waves, and we would be wise to remember that.

Using Lessons from the Past to Shape Our Future

Examples of the technologies we use daily and their impacts on our larger world can guide us as we explore how and when to use new technologies. The rapid development of technologies leading up to and throughout the twentieth century has been captured through academic study and historical records. Globally, we are blessed to have a rich body of knowledge about the diffusion of technology and a well-documented history of technologies since the Industrial Revolution. Unlike previous generations, we can use myriad technologies like metadata, optical character recognition (OCR), and online access to millions of documents to assess and prepare for emerging technologies. For the first time in human history, we have limitless potential to contemplate the future by assessing the past in sharp detail.

While we have so much to learn about the human mind and human behavior, we do know more about stimuli, responses, and basic human psychology than ever before. We know how humans adopt technology as a group, and we also know how humans use technology as individuals. This is another rich source of information we can use to guide us today as we explore the new innovations poised to impact our lives tomorrow.

For the first time in human history, we have limitless potential to contemplate the future by assessing the past in sharp detail.

Consider that every nation has aviation regulations that are based solely on the mechanics of flight as well as aviation regulations that are based solely on the behaviors of humans. Likewise, we would expect decisions and regulations around any new technology to take both aspects—the nature of the technology and the nature of human beings—into account. No conversation about how we adopt and adapt to technology should exclude the nature of the human beings who will use it. Technology does not exist for technology's sake. It exists for humans to use. So we have to learn to use it wisely.

DDT, mustard gas, Agent Orange, BPA, asbestos insulation, and lead paint all contain chemicals with terrible side effects on human beings. The hard lessons learned from these toxins, if we are wise enough to learn them, can provide the foresight to modify regulation and improve research and manufacturing processes. We now know better how to assess the potential impacts of chemicals before they find their ways into farms, warfare, or our homes. The lessons of the past have taught us to appreciate the risks of rushing into the next innovation without critical thinking and an honest appraisal of potential consequences. Moving forward, we owe it to the people of both the past and the future to do better.

No one is doomed to repeat the past. Through positive intention and direct intervention, we can choose a better path and build a meaningful legacy for future generations.

—VILAS DHAR

THE IMPACT OF MODERN TECHNOLOGIES

Not very long ago, unless you invested in an encyclopedia set, looking up an obscure fact might have required a trip to the library. If the library were closed, the bookstore was your next best bet. Even if you could order that book at midnight on one of the founding e-commerce sites in the mid-1990s, you wouldn't get it instantly.

In the early days of the internet, online transactions and shipping were much slower than we have now come to expect. Information was sparse, and its reliability was suspicious at best. We had no AI personal assistants to answer our questions or tell us the weather. It is amazing how quickly we adopt, adapt, and become dependent on technology in modern times.

Over the past twenty-five years, we've gone from professors disallowing citations from internet resources to students doing all their work on the cloud. Twenty-five years ago, even if you were allowed to cite an internet resource, many manuals of style wouldn't have complete methods for capturing it. Now several research sites provide free applets to export properly formatted citations for materials found in whatever formats are required.

This example is highly academic both in framing and content, but it also took more than two decades to transition from paper to applets. Faster things have happened when bigger money was involved. E-readers and social media were available in 2007, years ahead of more academic widgets, because money could be made.

> **Innovators in any technology are going to hit stumbling blocks over the infrastructure limitations they don't yet know exist.**

Early proponents of the World Wide Web hoped that like television before it, it would be a great equalizer, allowing information to be shared equitably with all. However, like television, the internet was its own kind of boomtown for commerce and moneymaking. One problem with that, as we saw with the first technology bubble in 2000, was that people had big dreams for using the internet but lacked the means to use it as quickly as humans could imagine new uses. Prospectors were showing up at the gold rush without any shovels.

Innovators in any technology are going to hit stumbling blocks over the infrastructure limitations they don't yet know exist. This has changed in more recent years, especially after the emergence of broadband and content streaming. Once the necessary infrastructure was in place to solve the problems early innovators didn't expect, much less understand, things became very different.

Consider the activity of ordering pet food from a specialty vendor. This smaller vendor now leverages the same supply chains and delivery services set in place by e-commerce giants. Our pet food example is hugely successful in 2022. However, the same business model with the very same concept went under in the year 2000. Back in 1999, getting pet supplies online was a cool idea that didn't work terribly well because in 2000, shipping was incredibly expensive and somewhat limited. Back then, inventory systems were not as sophisticated as they are today.

In 2020, at the height of the global pandemic, ordering pet food online could be literally lifesaving for humans and pets. In 2022, getting pet food online has simply become a habit, an adaptation of human behavior. Timing really is everything.

As explained in the previous chapter, until a technology or innovation is adopted by the masses, it is speculative at best. With many modern technologies, utility follows commercial opportunity. The ability to cut time out of processes of commerce habituates humans to instant gratification.

If we have everything we need, is necessity still the mother of invention, or have convenience or instant gratification become invention's new stepmom? What happens if you invent something, but it's already old before anyone can use it?

The rate of innovation and replacement shows no sign of slowing down, much less stopping. When we truly believe we need the newest gadget or our older objects fail due to planned obsolescence, we leave a trail of abandoned things in our wake. No one wants to go backward, but now it is time to pause and consider how to proceed conscientiously, knowing what we know now.

If you play the sound of a modem dialing up the internet to children born in the twenty-first century, they will have no visceral reaction to it. Most of them won't even recognize it. However, for early adopters and an early majority of internet users, memories of waiting, and waiting, and waiting for an image to load on a web page will be conjured by that sound. We are human, with ancient neural pathways that react to stimuli, even stimuli long forgotten.

No one wants to go backward, but now it is time to pause and consider how to proceed conscientiously, knowing what we know now.

Vinyl records and Polaroid pictures are fun retro trends having second lives, but at one point in the still-living memories of some, they were novel. Before the comeback of the LP, children in the 1990s might have asked what a record was, only to be told it was like an old compact disc (CD). Just ten years later, children didn't know what a CD was, so they might be told that an LP and CD are both like a tiny iPod. In another ten years, the iPod, CD, and LP would all be equated with a music-streaming icon. This is a revolution that is much shorter than a generation. Siblings under the same roof will have different formative contexts for technology.

Once, owning a video cassette recorder (VCR) was life changing; then Netflix early adopters were returning those red envelopes at their mailboxes. Now everyone knows how to stream movies. Every generation is forced to adopt and adapt quickly. The pace of our changing world would be exhausting if it were not so stimulating.

The downside to the rapidity of change is the high cost to

consumers and to the earth. How do businesses and schools manage technology budgets and replacement strategies if computers and phones must cycle out every two years to keep up? Manufacturers are intentionally evolving hardware and software configurations to force consumers to adopt newer and newer technologies. The side effect of this lucrative business proposition is postconsumer waste.

> **Every generation is forced to adopt and adapt quickly. The pace of our changing world would be exhausting if it were not so stimulating.**

According to 2019 *TIME* magazine article, "Less than a quarter of all US electronic waste is recycled, according to a United Nations estimate. The rest is incinerated or ends up in landfills. That's bad news, as e-waste can contain harmful materials like mercury and beryllium that pose environmental risks."[3] This is a problem on a global scale.

The Pendulum

Even as the list of unintended consequences from the explosion of modern technology can seem bleak, the same technology has benefited us in ways our predecessors could have never imagined and on which our successors will depend. Therefore, it is essential for us to take stock of where we stand. Our modern technology has made human reunions and scientific breakthroughs possible, and it has also enabled crimes and attacks against individuals, including children, businesses, charities, entire nations, and nature itself. Here is a limited sample set of considerations we need to contemplate with current and emerging technologies in 2022.

---

3 Alana Semuels, "The World Has an E-Waste Problem," *Time*, June 3, 2019, https://time.com/5594380/world-electronic-waste-problem/.

Social media. Social media arrived in most of our lives in the beginning of the twenty-first century, and while we are edging ever closer to the second quarter of this century, it still feels new in many ways to all, even those born into the digital landscape. Video conferencing and direct video calls made it possible for our society and economies, though bruised, to survive a global pandemic.

Classrooms and businesses kept going, while people who were near death were able to hear the voices and see the faces of their loved ones on devices when they could not be together in person. Newborn babies who could not be seen by grandparents in person during the darkest days of the global pandemic could be seen and heard by grandparents, aunts, uncles, and big brothers and sisters until times improved. These technologies are genuinely life changing. While none of this was or is a direct replacement for human connection, it was better than the utter absence of connection we would have experienced without those technologies.

On the flip side, the instant gratification we experience through social media that gives us unfettered access to and creates new kinds of celebrity can spark a deep sense of inadequacy in children, adolescents, and young adults. In the worst cases, this can lead to eating disorders and self-harm. Streaming media on demand compounds our collective need for instant gratification.

Though these troubles are not limited to children, children *are* disproportionately influenced by unchecked *influencers* and impacted by cyberbullying due to their underdeveloped social skills and the amount of technology they consume. At a minimum, cyberbullying can leave victims feeling isolated and depressed. At its worst, cyberbullying can be the tipping point that drives a victim to take their own life. A 2022 *Journal of the American Medical Association* article called for pediatricians to ask their patients about exposure to cyberbullying, as it is a direct indicator for suicidal ideation.[4]

Crimes. The ever-increasing accuracy of DNA analysis, aided by powerful computers, advanced scientific study, and the public's willingness to share their own genetic fingerprints to help investigation, has led to the resolution of several murders that happened more than forty years in the past. These breakthroughs are delivering overdue justice to families and law enforcement experts who never could have hoped for resolution before 2018. Likewise, the proliferation of CCTV and doorbell cameras has led to faster apprehension of criminals and acts as a deterrent to some criminals. These innovations offer society hope and confidence. While privacy concerns are as much an issue as when fingerprints and blood typing were first used, for society the debate will continue, even as technology advances.

On the digital front, the threats against electronic security seem endless. From identity theft, to hacking and denial of service (DNS) attacks, to literal terrorism, cybersecurity is one area of technology that can never be overfunded or overstaffed. According to the World Economic Forum:

> In 2020, malware and ransomware attacks increased by 358% and 435% respectively—and are outpacing societies' ability to effectively prevent or respond to them. Lower barriers to entry for cyberthreat actors, more aggressive attack methods, a dearth of cybersecurity professionals and patchwork governance mechanisms are all aggravating the risk.

With this exponential rise in cybercrime, we need to look long and hard at where we stand today before we leap into new technological frontiers.

---

4 Shay Arnon et al., "Association of Cyberbullying Experiences and Perpetration with Suicidality in Early Adolescence," *AMA Network Open* 5, no. 6, 2022, https://jamanetwork.com/journals/jamanetworkopen/fullarticle/2793627.

Marketplaces. The evolution of technology and cloud computing has made e-commerce more accessible to individual creators and small businesses alike. It is now possible for people to engage in passive income markets and the gig economy. Self-publishing, content creation, and several task-oriented job boards are avenues for revenue for millions of people.

Artists and crafters have several means to sell their wares online, creating markets in places that they could never access in their own communities. Additionally, they can tap into ready-made shipping and marketing platforms meant for individual creators. The potential for individuals to follow their passions or pursue more income is limited only by the amount of time they have available to invest in the work.

Unfortunately, this same technology has established an entire supply chain for human trafficking. Predators can find, approach, and groom potential victims through social media. Then they can use the same tools you might use to sell your old yard furniture to sell human beings for sex or other forms of enslaved labor through on the dark web. The internet has taken the place of the seedy street corner for those who traffic in human beings while increasing the risks to victims and decreasing the risks to perpetrators through anonymity.[5, 6]

The material world. Cell phones are made sleeker and smarter with every iteration. They have gone from being simply cordless telephones to being phones, cameras, video cameras, address books, personal

5 Working Group on Trafficking in Persons, "Conference of the Parties to the United Nations Convention against Transnational Organized Crime," *United Nations*, October 12–13, 2021, https://www.unodc.org/documents/treaties/WG_TiP_2021/CTOC_COP_WG.4_2021_2/ctoc_cop_wg.4_2021_2_E.pdf

6 Judge Herbert B. Dixon Jr., "Human Trafficking and the Internet* (*And Other Technologies, too)," *The Judges' Journal*, Winter 2013, https://www.americanbar.org/content/dam/aba/publications/judges_journal/vol52no1-jj2013-tech.pdf.

computers, flashlights, and safety beacons. The smartphone is both a tool and a defensive weapon that can give us some degree of comfort and security. It keeps us connected, and with some features, we can keep track of our children and be found if our cars skid off the road.

Yet imagine going through cars as quickly as we go through telephones or even televisions. Excessive postconsumer waste is the endpoint of a global manufacturing craze that begins by digging up the earth for copper and other elements. We dig deep to create treasure that becomes trash in the same decade. We are burying ourselves alive in waste—and that's when we don't burn it, sending toxic exhaust into the atmosphere.[7] We might consider cleaning up our act before creating more and newer cycles of mass production and trash.

The Lessons

The World Wide Web was called the internet superhighway in its earliest days, and that analogy holds when we think about regulation and safety today. The department of transportation does have agencies devoted to safety. On the contrary, there are not any organizations dedicated to internet safety.

Since 1999, the following regulations have been implemented in the United States:

- 1999: The Anticybersquatting Consumer Protection Act (ACPA)[8] to prevent the hoarding of web addresses

7 Syed Faraz Ahmed, "The Global Cost of Electronic Waste," *The Atlantic*, September 29, 2016, https://www.theatlantic.com/technology/archive/2016/09/the-global-cost-of-electronic-waste/502019/.

8 106th Congress of the United States of America, *S.1255, Act of Congress*, August 5, 1999, https://www.congress.gov/106/bills/s1255/BILLS-106s1255es.pdf.

- 1999: The Gramm-Leach-Bliley Act (GLB ACT)[9] to ensure that financial institutions let consumers know how their private information is used
- 2000: The Children's Internet Protection Act (CIPA)[10] to prevent the viewing of sensitive content in schools and libraries
- 2000: The Children's Online Privacy Protection Act (COPPA) to prevent collecting information about and restricting access to children over the age of thirteen
- 2006: Internet Freedom and Nondiscrimination Act[11] to prevent discrimination by and monopolization by internet service providers (ISPs)
- 2006: Federal Trade Commission Act—US SAFE WEB Act[12] amendments to counter spam, spyware, and fraud

According to the *New York Times*, "the United States doesn't have a singular law that covers the privacy of all types of data. Instead, it has a mix of laws that go by acronyms like HIPAA, FCRA, FERPA, GLBA, ECPA, COPPA, and VPPA."[13] Many of the laws cited by the *Times* article are more than twenty years old. Considering the speed

---

9 Federal Trade Commission, "Gramm-Leach-Bliley Act," May 2022, https://www.ftc.gov/business-guidance/resources/ftc-safeguards-rule-what-your-business-needs-know.

10 Federal Trade Commission, "Consumer Guide: Children's Internet Protection Act," December 30, 2019, https://www.fcc.gov/sites/default/files/childrens_internet_protection_act_cipa.pdf.

11 109th Congress of the United States of America, *H. R. 5417 [Report No. 109–541]*, June 29, 2006, https://www.congress.gov/bill/109th-congress/house-bill/5417/text.

12 109th Congress of the United States of America, *Public Law 109-455*, December 22, 2006, http://uscode.house.gov/statutes/pl/109/455.pdf.

13 Thorin Klosowski, "The State of Consumer Data Privacy Laws in the US (And Why It Matters)," September 6, 2021, https://www.nytimes.com/wirecutter/blog/state-of-privacy-laws-in-us/.

at which modern technology evolves, it seems lagging and lacking to have so little regulation and oversight on topics including cyberbullying, exploitation, cryptography, and internet privacy and no regulations on postconsumer electronic waste. We have a law to protect children from seeing pornography on the internet but too few laws to protect children from being victimized through the same technologies.

All our contemporary shortcomings beg the question: As a society, are we socially ready for the next wave of technology? Probably not, but the next wave is coming whether we are ready or not. When it comes, we will benefit as much as we will struggle, but we will sell ourselves short if we ignore the lessons of our past as we face the future. So let us prepare ourselves.

As a society, are we socially ready for the next wave of technology?

Technology drives up the rate of change and availability of data, making our decisions more complex. Your legacy is how you teach your teams to make decisions in this environment.

—JANA EGGERS

PART TWO

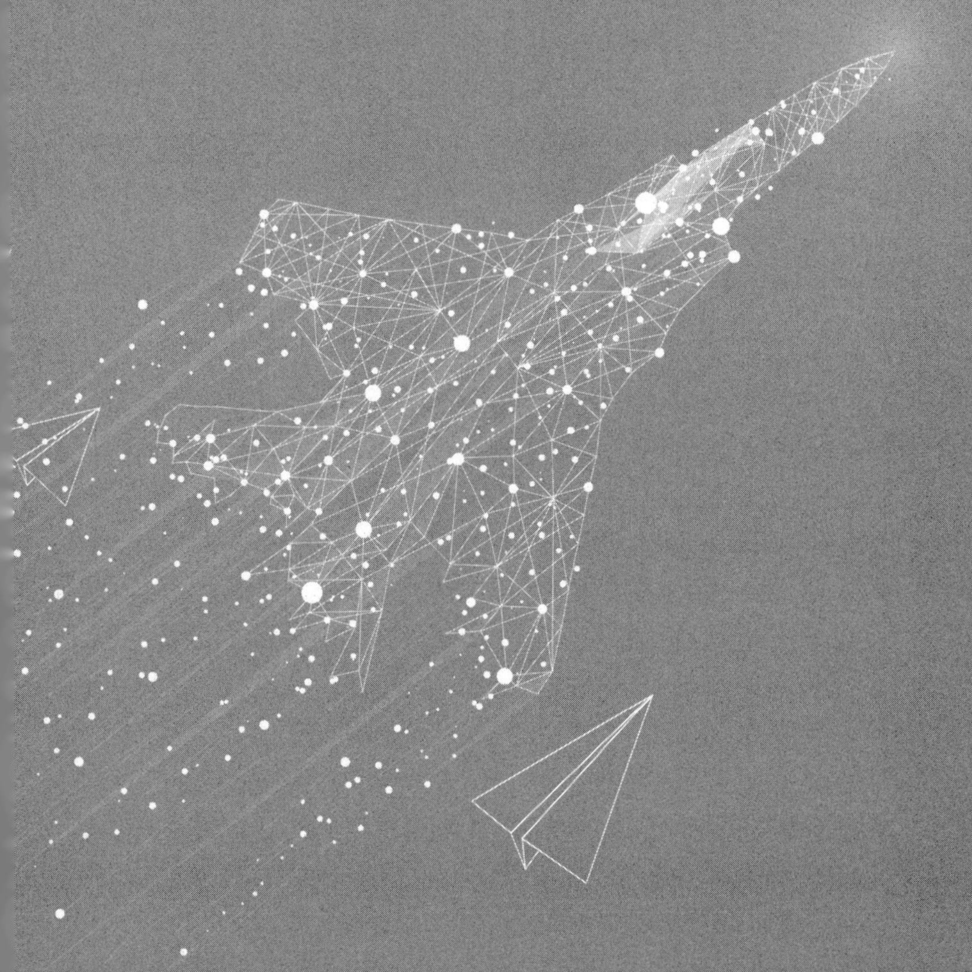

Research Lab
Estimated time for processing: 90 days,15 hours
1996
Estimated time for processing: 3 days, 6 hours
2026

QUANTUM COMPUTING

The director of BAB pharmaceutical research, Dr. Akpofure Efe, inherited a major problem when he took over leadership. One of BAB's best pain medicines called Releze had unintended side effects in women of childbearing age. These effects included memory loss, dizziness, and occasional fainting. In more extreme cases, liver inflammation was detected. Releze never had statistically significant side effects like these found in the general population in the studies done before the drug went to market. Dr. Efe needed to understand what was happening and how they had missed it in the clinical trials.

Through hundreds of hours of analysis, Dr. Efe's team discovered that women metabolized the drug far more slowly than men did. Because the original clinical trial participants were primarily college-aged men, too few women had been in the clinical trials to detect the statistical significance of Releze on female patients. The problem was prevalent because Releze, being a nonopioid and nonaddictive pain reliever, was prescribed widely to women in postnatal care. This phenomenon was how the side effects came to light. Postnatal caregivers were prescribing Releze to be taken every four hours, but Dr.

Efe's newer studies showed that women should take it every six hours, given the buildup in their bloodstreams and livers. It turned out, with more study, that not only women of childbearing age were affected; however, they were the ones most likely to take the drug and be in medically supervised situations wherein side effects were reported.

Now that they knew the ramifications, Dr. Efe had to take immediate action. He worked with the BAB legal and marketing teams to get ahead of the problem and in turn worked with different government agencies in the areas where Releze was sold to change the recommended doses for women.

After the urgency of the Releze overdosing problem was addressed, Dr. Efe sought out Dr. Sriji Sankar, PhD, the BAB director of emerging technology, to discuss methods they could use to prevent another Releze situation by evaluating their existing drugs against the metabolic information they had discovered with the recent emergency.

Was there a way BAB could compute the scale of these problems across the entire population, including genetic aspects, such as the presence or absence of a Y chromosome? Could the women impacted by the use of Releze and other drugs be saved the consequences of a perpetual state of excessive dosing? Might accidents be prevented? Lives saved?

While Sriji offered some consolation to Akpofure with data analysis they could perform right now, she held out hope that they could prevent new drugs from having the same issues in the future through the possibility of quantum computing.

Understanding Quantum Computing

Simply stated, quantum computing uses qubits instead of the traditional bits that our classical computing systems currently use. Bits are binary; they can be 0, or they can be 1, but they cannot be both. Qubits can be both 0 and 1, and their value is determined based on their purpose (superposition) and interactions (entanglement) while processing information. Qubits are fluid in application and will process data infinitely more quickly than bits because all qubits share that fluidity. Acting without binary constraints, all possible paths can be explored by quantum computers.

When scientists need to develop new drugs and chemicals, they often need to examine the exact structures of molecules to determine their properties and understand how they might interact with other molecules. Unfortunately, even relatively small molecules are extremely difficult to model accurately using classical computers, since each atom interacts in complex ways with other atoms. Currently, it's almost impossible for today's computers to simulate basic molecules that have relatively few atoms.

Quantum computing's potential lies in its exponential power to manage and manipulate data and information at accelerated speed to great advantage. It is in that accelerated computation that quantum computing has the most promising aspects for solving some of the world's most complex problems. However, without foresight and oversight, that same technology has the potential to exacerbate our biggest social problems and create new problems we cannot yet imagine.

> **Quantum computing's potential lies in its exponential power to manage and manipulate data and information at accelerated speed to great advantage.**

Quantum computing, still in its infancy, will ultimately be capable of deriving computations for variables in an infinite pattern of multiple states, performing thorough predictive analysis and modeling. This means complex problems that require multivariate analysis of factors in a state of change can be solved through modeling those factors throughout the many dimensions of change.

With superposition, we will be able to "hit" moving targets like epigenetics that are affected by heredity, climate, environment, exposure, and hundreds of other factors that cannot be controlled. It's the need for controls that limits the predictability of the existing body of knowledge in most sciences. By eliminating the need to analyze single states sequentially, comparative analysis offers endless possibilities.

THE OPPORTUNITIES

Quantum computing harnesses the laws of quantum mechanics to solve problems too complex for other computers. It draws on the hardware and software legacies of computer science against the infinite possibilities of physics.

The potential of quantum computing to accelerate drug development to reduce adverse reactions and overdosing through predictive analysis is, of course, a beautiful idea. Imagine using the data to determine minimum viable dosage and prescription surveillance in such a way that we could avoid the successor of the opioid epidemic. Imagine the ability to better predict the season strains of influenza for better vaccination planning. The potential for human good is as exponential as the power of the qubit.

> **The potential for human good is as exponential as the power of the qubit.**

Quantum computing has potential applications in material sciences, medicine, drugs, genetics, military intelligence and warfare, artificial intelligence, communication, economics, transportation, and countless other domains. Essentially, any discipline that impacts humanity can either be in the crosshairs of quantum computing or can benefit from it. For this reason, it is crucial to understand its risks, implications, and potential unintended consequences.

THE CHALLENGES

Unlike innovations like social media and the internet, we are advantaged to be on the edge of the global quantum computing frontier. Because humanity has navigated innovation and diffusion of technologies repeatedly since Gutenberg's press sparked the rise of Protestantism to the use of chemical warfare in the trenches of WWI, we know better than to allow a technology to get out of hand. The question is: What will we do to keep it focused and bounded from dangerous exploitation, monopolization, and abuse before that genie is out of the bottle?

The question all of society must ask about this potential is *whether* that advantage will benefit all of humanity or exponentially widen the already gaping divide in equity and wealth. Thought leaders in quantum computing represent an expanding cross section of multiple domains, including philosophy, physics, computer science, business, medicine, and economics. This diversity of global stakeholders is not only drawn to the potential power of quantum computing but also to the nature of the technology.

Quantum computing will rely on algorithms that will evolve without human interaction. This means that the analysis and predictions computed in quantum processing will be done in a true "black box." Imagine that the computer gives us an answer, but we cannot see the work or validate it. It's not only possible but also likely that the computer is right. But what if it's not? How do we trust what we cannot see? How can we develop faith in the machine?

Quantum computing has the potential to decrypt all our best encryption standards, which will put us in a permanent state of catch-up for cybersecurity. How can we prevent hacking if the hacker is a quantum computer? Existing risks to sovereign voting rights, identity theft, fraud, and internet terrorism are all posed to escalate with the rise of quantum computing. How will we prepare for a world where a single computer can steal an election? How do we preserve voters' rights when a single bad actor with the right technology can nullify the electorate?

Outside of the risks quantum computers themselves pose, the risk of exploitation is another concern. What good is a cure for cancer if no one can access it because the company with the cure makes it cost prohibitive? Likewise, the financial volatility of markets already serves those with the power to exploit existing information. What happens when those same people control modeling that allows them to game the system with impunity because they own the machines that make it possible?

What about human anatomy? If we know that a particular disease is hereditary and quantum computing tells us that a certain group of people will carry that disease forward as carriers, even if they disease doesn't exist in their present offspring, how does society apply that knowledge? Today, carriers and potential carriers of certain hereditary diseases (like cystic fibrosis and muscular dystrophy) can receive genetic counseling along with genetic testing to make informed decisions. What happens when the data is coming faster than the protections to help people understand that data and what it means to them?

What happens when the data is coming faster than the protections to help people understand that data and what it means to them?

We are already seeing what DNA databases have done revealing relations and uncomfortable family secrets and solving crimes that are several decades old. As that data becomes richer and more plentiful, how do we both preserve privacy *and* protect people from being traumatized by the fruit of this new tree of knowledge? Who will be the new gatekeepers of the quantum realm, and what will the world look like if there are no gatekeepers? How will we balance the potential for good against the potential for harm in a world where we are tempted to believe that a computer has all the answers?

The use of quantum computing will advance technology in ways previously thought unknown. However, we will in turn need to ask how it will remain ethical and consider what actions will need to be taken to accomplish that.

USE CASE ONE: OPTIMIZED COMPUTATION AND ANALYSIS OF DATA

Though quantum computing is still hypothetical, we do know that it will allow for data to be studied and analyzed in an optimal manner. Some possible use cases include testing for infinite interactions between drugs or optimizing supply chain in such a way that virtually all possible scenarios can be analyzed and the most efficient path chosen. In this scenario, we must ask the following:

- How will user data remain secure and the privacy of the individual or entity be respected?
- What risks exist for people and entities whose data is being leveraged?
- How will the accuracy of new quantum algorithms and their subsequent results be verified and ensured?

- Will businesses be able to leverage these algorithms to their advantage (e.g., what will the learning process and accessibility look like)?

USE CASE TWO: ENCRYPTION AND DECRYPTION

Quantum computing will allow for the factoring of larger and larger prime numbers. Even the strongest cryptographic code currently known will be breakable faster than ever.

- How will encryption methodologies change to ensure the secure storage of information and keep pace with hackers using the same technologies?
- Cyberattacks could very likely become more prevalent, so how will regulation and policy change to prevent and combat cyberattacks?
- How will we deal with the amount of memory required for such powerful computations while considering efficient and sustainable resource allocation?

USE CASE THREE: CYBERSECURITY AND DATA PROTECTION

It is no secret that the cybersecurity industry has grown from the need to protect and secure data. With the potential rise in the use of quantum computing, there is a huge threat to deencrypt data.

Quantum computers will provide hackers with uncontested paths to all private data. This could leave many people at risk of identity theft, bank account theft, or even the release of state secrets, as well as many other privacy breaches.

- How can organizations effectively mitigate this type of risk and lay the groundwork for an ethical rollout of quantum computing at scale?
- If stolen encrypted data is later released, who would be held accountable? Many individuals will be affected, but would a single person or organization be directly liable?

Sample Questions for Further Conversation

- How will organizations ensure the accuracy of results produced by quantum computing?
- What will be the probabilistic threshold or margin of error for "correctness," and who is responsible for determining thresholds?
- How will developers and the public be able to trust the reliability of algorithms?
- Will quantum algorithms follow a logic that can be explained and traced?
- How will debugging quantum software be relevant, given that it will employ algorithms and computing far more complex than ever encountered before?

- Will developers be kept accountable for the transparency of their algorithms and their reliability?
- How will organizations ensure that users are aware of margins of error and the probabilistic manner of certain quantum software?
- What risks will quantum computing pose to secure and encrypted user data?
- Quantum computers will allow for the breaking of any presently known cryptographic code, so how will encryption techniques change to secure information?
- What will be the implications of quantum computing and decryption on secure government data and international relations?
- Should the internet begin moving toward a new form of encryption? If yes, when should this shift begin?
- What safeguards and policies will be put in place to protect user data?
- Due to the complexity of quantum algorithms, will user-friendliness ever be a possibility?
- How will the data fed into quantum models be regulated to ensure fairness and eliminate biases? Will responsibility fall into government regulation, or should it be up to companies?
- Possessing knowledge of quantum computing will provide individuals and entities with power, so how will this centralization of power impact current inequities we see in the world with wealth, social opportunity, etc.?
- Will developers be employing quantum algorithms and

models for social good?

- Will the development of quantum computing ultimately prove beneficial to humankind?
- For specific quantum software, who is to be held accountable for errors and disruptions?
- Does the organization have an established protocol to follow if something goes wrong?
- Were key individuals from groups or departments across the organization (e.g., technology, operations, talent, etc.) included in designing and implementing quantum algorithms?

As our physical and virtual worlds grow more blended, let's constantly ask ourselves what attributes from the former we want to leave behind or bring forward into the latter.

—ATUL PATIL

NFT
accepted here
Urgent Care

NFT
accepted here
Saloon

BLOCKCHAINS, NFTS, AND CRYPTOCURRENCIES

Hector Reyes, BAB's chief financial officer, was recently approached by Phuong Le from the Vietnam development office. Phuong was coordinating a vaccination program and told Hector that many of the young people they wanted to enroll in their vaccination clinics for clinical trials prefer to be paid with cryptocurrency, but BAB policy neither includes nor excludes cryptocurrency.

Hector reached out to Dr. Sriji Sankar, the director of emerging technologies, to get her take on cryptocurrency and when and how BAB should enter this uncharted water as a global company. Sriji explained that Vietnam, especially its youth, was leading the world in cryptocurrency adoption. She recommended making Vietnamese clinical trials case studies on how to use cryptocurrency for paying study participants who opt in to that payment method. Her team helped Hector and Phuong set up and monitor the program.

If the program is a success, BAB will be seen as an industry leader in adopting cryptocurrency and recognizing the needs of their study participants. Everyone on the BAB case-study team is excited to see what will come of this project.

Understanding Blockchains, Cryptocurrencies, and NFTs

A simpler way to understand the complex individual parts of blockchains, cryptocurrencies, and nonfungible tokens (NFTs) is to see them as comprising a system. Blockchains are infrastructures or networks of nodes that support both cryptocurrencies and NFTs, neither of which can exist without blockchain. The core purpose of blockchains is to host an encrypted distributed ledger that is a single source of truth for transactions held by multiple parties or nodes to ensure that accuracy is not compromised through accidental influences or the abuse of control from a central party.

The concept of the distributed ledger is so significant to blockchains that blockchain systems are also known as distributed ledger technology (DLT). The World Bank defines this concept:

> Blockchain or DLT are the building blocks of "internet of value," and enable recording of interactions and transfer "value" peer-to-peer, without a need for a centrally coordinating entity. "Value" refers to any record of ownership of asset—for example, money, securities, land titles—and also ownership of specific information like identity, health information and other personal data.[14]

14 The World Bank, "Blockchain & Distributed Ledger Technology (DLT)," *Understanding Poverty Web Content,* April 12, 2018, https://www.worldbank.org/en/topic/financialsector/brief/blockchain-dlt#:~:text=Distributed%20ledgers%20use%20independent%20computers,in%20an%20append%20only%20mode.

This concept of value touches on what cryptocurrencies and NFTs mean in relationship to blockchains. To simplify this further, consider that both cryptocurrencies and NFTs are simply tokens. Tokens are discrete and traceable pieces of information held within the distributed ledger of the blockchain in which they exist and are traded. The difference between cryptocurrencies and NFTs is the term *fungible*. *Fungible* simply means "interchangeable." For instance, any two US pennies in common circulation have the same value. These pennies are fungible items within a system of currency. The same is true with cryptocurrency tokens; any one unit is interchangeable with another.

Consider that both cryptocurrencies and NFTs are simply tokens. Tokens are discrete and traceable pieces of information held within the distributed ledger of the blockchain in which they exist and are traded.

However, unlike a penny or a dollar bill, the transactions made with cryptocurrencies leave discernible trails in blockchains. This allows fungible assets to be tracked back to the source to ensure all transactions are valid through the distributed controls of the blockchain. Still, the units of currency themselves are not relevant; rather, the transactions are what is traced. This is very different with nonfungible tokens.

Nonfungible means that these tokens are unique and not interchangeable. An NFT is a unique cryptocurrency token that can take the form of anything digital—a drawing, music, a video, a meme, or Jack Dorsey's first tweet. So NFTs allow you to buy and sell ownership of unique digital items and keep track of who owns them using blockchains. The current market for NFTs is centered on the fine art marketplace and on internet-based games in which in-game activity can lead to the acquisition of NFTs. In the case of fine art, the token in the blockchain is representative of the artwork, not the artwork itself. Most often this is digital artwork, but it could be representative of three-dimensional art. Blockchains allow users to certify and indemnify secured ownership of irreplaceable and unique items to prevent theft by fraud and to recover and reestablish ownership in cases of theft by other means.

Thus, blockchains allow users to conduct transactions and establish certified collections with both fungible tokens (cryptocurrencies) and nonfungible tokens (NFTs). This is all fine, but what makes these better or worse than our current forms of transactions? Why would we want to use these technologies? What good are they?

THE OPPORTUNITIES

While it may presently be easier to buy local items with cash or by tapping a credit card, the evolution of cryptocurrency may have a greater influence in the global marketplace. Presently, brokers and other internet-based intermediaries control transactions in the gig economy. Because these businesses are global in nature and can be used by both travelers and individuals who want to hire contract labor overseas, they generate their income on transaction fees.

Banking and supply chain transactions could benefit from more secure and traceable routing. There is extraordinary potential in blockchains for these industries.

If those transaction fees were removed along with intermediaries and replaced by a global currency—any cryptocurrency that can be used globally—then the gig economy could be more advantageous to both the buyers and sellers of services. Transaction fees will still exist in blockchains, but like the ledger, they are distributed, and these fees are limited to the cost of operation wherein the individual nodes (and their owners) that validate and certify transactions are paid out a proportion for their technical contribution. However, this cost is currently less than the costs of using intermediaries. With the proliferation of blockchain systems and ever-increasing computing power, the cost of blockchain verification should not outpace the transaction fees of gig-economy intermediaries.

However, just like any currency, cryptocurrency is not immune to inflation. Furthermore, until it is ubiquitous enough to buy commodities and necessities, the hoops that holders of cryptocurrency must jump through to make their tokens as good as cash is a barrier to entry for many individuals who might otherwise be interested in joining a blockchain network.

THE CHALLENGES

Still, the technology that enables NFTs in blockchains is appealing for markets like real estate transactions. From condominium ownership to reverse mortgages, using NFTs for title search and validating transactions of real property holds great appeal. Similarly, banking and supply chain transactions could benefit from more secure and traceable routing. There is extraordinary potential in blockchains for these industries, and banking is already leading the way with the expansion of these technologies.

To understand the power of cryptocurrency to attract bad actors, we do not need to speculate because we can look back a decade. When Ross William Ulbricht, an American graduate student, founded his dark web site, the Silk Road, in early 2011, his intention was to use anonymity to allow individuals to buy and sell drugs without consequence. While this was hardly a noble effort, the unintended consequences of the Silk Road went far beyond individual drug trades. Ulbricht capitalized on the emerging cryptocurrency Bitcoin, which relies on blockchains, to enable the financial part of creating "untraceable" illegal transactions over the internet, as in 2011 it was thought that Bitcoin was anonymous.

As Bitcoin did not collect social security numbers or bank accounts to open new user accounts, the single requirement of a mailing address could be obfuscated with a fake or borrowed address of a post office box. While this was a valid assumption with what was known in 2011, Ulbricht and his customers would ultimately learn that was not a guarantee of anonymity or impunity.

During its thirty-two-month existence, Silk Road earned Ulbricht over eighty million dollars in commissions, which was only 10 percent of the flow of transactions through the site. In its last months of operation, Ulbricht was taking in over ten thousand dollars each day in commissions. Though he experienced hacks, blackmail, and lost revenue, Silk Road was still highly profitable for Ulbricht and his many customers. In fact, Silk Road processed over 9.5 million Bitcoins of the 11.75 million in circulation between 2011 and 2013, meaning most of the Bitcoin currency in circulation was used for illegal transactions.[15]

Even though the site existed on the dark web, Ulbricht still needed a customer base, and he was proud of his site. His efforts to build Silk Road caught the attention of the press and of law enforcement, leading to his eventual downfall. The sting operation and subsequent charges were the stuff of books and movies—quite literally, no fewer than six books have been written on Silk Road since Ulbricht's arrest. However, the broader social implications for a marketplace that seeks to usurp laws is best expressed by the US attorney Preet Bharara, who successfully prosecuted Ulbricht:

---

15 The United States Department of Justice, "Ross Ulbricht, A/K/A 'Dread Pirate Roberts,' Sentenced In Manhattan Federal Court To Life In Prison," *The United States Attorney's Office Southern District of New York,* May 29, 2015, https://www.justice.gov/usao-sdny/pr/ross-ulbricht-aka-dread-pirate-roberts-sentenced-manhattan-federal-court-life-prison.

"Make no mistake: Ulbricht was a drug dealer and criminal profiteer who exploited people's addictions and contributed to the deaths of at least six young people. Ulbricht went from hiding his cybercrime identity to becoming the face of cybercrime and as today's sentence proves, no one is above the law."[16]

Ironically, Bitcoin explicitly states that it is not anonymous on its website: "Bitcoin is designed to allow its users to send and receive payments with an acceptable level of privacy as well as any other form of money. However, Bitcoin is not anonymous and cannot offer the same level of privacy as cash. The use of Bitcoin leaves extensive public records."[17] It is the very nature of blockchains—which support cryptocurrencies, including Bitcoin—that creates those extensive public records. Those same records enabled the US government to seize one billion dollars' worth of Bitcoin from a single hacker in 2020, seven years after they hacked Silk Road.

It is crucial for society to understand technologies as they emerge and before they become commonplace.

Every technology is rife for exploitation, which is why it is crucial for society to understand technologies as they emerge and before they become commonplace. Even if we don't yet use them, it is important to know what blockchain and the associated technologies it supports like cryptocurrencies and nonfungible tokens (NFTs) mean to the social order and what opportunities and implications they pose for our collective future.

With the opioid epidemic that erupted in the United States during the same decade as Silk Road, it is not hard to see why anonymity in

---

16 The United States Department of Justice, "Ross Ulbricht, A/K/A/ ' Dread Pirate Roberts,' Sentenced In Manhattan Federal Court to Life In Prison."

17 Bitcoin, "What Is Bitcoin?," *bitcoin.com,* retrieved December 22, 2022, https://BitCoin.org/en/faq#is-BitCoin-fully-virtual-and-immaterial.

cryptocurrencies and blockchains is simultaneously appealing and frightening. In the case of the dark web, where weapons of mass destruction, human trafficking, and money laundering are all very real cybertransactions, secure but traceable blockchains make sound ethical sense. When does the privacy and protection of personally identifiable information cross the line from making good sense to being dangerous?

According to Earthweb, there are four blockchain network types and over one thousand instances of blockchain platforms existing on the internet.[18] On top of these thousand blockchain platforms sit anywhere from twelve thousand to nineteen thousand cryptocurrencies.[19] This proliferation of cryptocurrencies is a danger for consumers whose investments in these markets could disappear overnight. While blockchains are meant to prevent fraud, the sheer volume of instances is a potential breeding ground for scams, especially because it is difficult to understand cryptocurrency, but it receives a great deal of hype.

This brings us to a third risk concern in the case of NFTs, where the implications are not quite so extreme or even illegal; still, they raise concerns. First is the carbon emission generated by NFT transactions, which is dependent on how much computing power is used in the methodology each blockchain platform uses to validate transactions. Because computing power is also distributed, there is no way to ensure or measure specific output of carbon emissions, but the more complex the validation, the more computing power and subsequent carbon emissions will increase. It is hoped that this will be curtailed in the future, but climate impacts are an omnipresent problem.

---

18 Thomas McGovern, "How Many Blockchains Are There In 2022?" *Earthweb.com*, September 11, 2022, https://earthweb.com/how-many-blockchains-are-there.

19 Arjun Kharpal, "Crypto Firms Say Thousands of Digital Currencies Will Collapse, Compare Market to Early Dotcom Days," *CNBC.com*, June 3, 2022, https://www.cnbc.com/2022/06/03/crypto-firms-say-thousands-of-digital-currencies-will-collapse.html.

Then there is *flex*, a term used to define the social capital of NFT ownership. That ownership may be a multimillion-dollar piece of digital art or a million-dollar tweet, but it is essentially a sign of great wealth. If NFTs serve no greater purpose other than to be status symbols of conspicuous consumption, what value do they really offer society?

Whether or not the minting of an NFT creates an asset that didn't exist before, every new NFT creates something else of value: scarcity. In the physical world, scarcity develops for well-understood reasons and seldom appears out of nowhere. NFTs mint scarcity from the electronic ether and lure consumers with their moneymaking potential. Most of us can remember a time when submitting our credit card information to an e-commerce site was considered foolhardy. Socially and economically, we have moved past that to a new arena of trust. Can the future of NFTs follow the same path and establish the basis for a new world of commerce and value?

Clearly, blockchains, cryptocurrencies, and NFTs are all innovations with great promise and equally great risks. If blockchains serve the purpose of democratizing value through a peer-to-peer marketplace of cryptocurrencies and NFTs and create a free market where intermediaries no longer profit from or gatekeep transactions, do we trust our peers? Money itself is a social construct. It is an idea and an abstraction, and the value of money is influenced by human emotions, as can be seen in every economic swing since the human invention of symbolic coins. So this abstraction has been taken to a new level on blockchains, and it brings with it all the same concerns of influence and equity we have had around money and economies for centuries.

The new concerns relate to the ease of transactions for criminals and the risk to victims. Humanity already has extraordinary challenges with human trafficking, crimes against children, and terror threats. How do blockchains exacerbate these problems? How can we influence them to curtail these problems instead?

How can we monitor these microcybereconomies to ensure that we are not destroying the tangible earth to power these intangible exchanges? What can we do as a society to protect the vulnerable from fraud, not only inside the blockchains but also in the misrepresentation of blockchains and cryptocurrencies? What controls can we take from the economies we currently use to ensure that we do not re-create the same exploitation and problems in these new economies?

CASE ONE: CREATING DECENTRALIZED MARKETPLACES

NFTs can represent various digital assets, such as art and virtual real estate. If someone buys one of these, then they own it and can sell it or rent it out as they wish.

Discussion:

- What safeguards exist or will be put in place to protect users' digital assets?
- How will marketplace actors be held accountable for fairness in transactions and investment?
- What section of the population will have access to this technology? Will access be equitable and inclusive?

CASE TWO: NFTS AS UTILITY TOKENS

An NFT can be used as a key to unlock access to a piece of content, allowing content creators to monetize their work in a new way.

Discussion:

- How will content creators and other vendors respond to security breaches? Who will be the point of contact for responding to breaches?
- Will the technology behind tokens be secure? Is it possible for someone to reproduce or reverse engineer tokens?

CASE THREE: SALES AND TRADING

Blockchains can create new possibilities to match investor demands. They can enable digital securities to efficiently go to market through mechanisms such as bilateral negotiations, matching algorithms, and decentralized exchanges. Digital security issuance is customizable and will thus enable more instantaneous and seamless business functions.

Discussion:

- Can marketplace regulation adapt to the increased use of blockchains in sales and trading?
- Will a lack of mediation lead to power imbalances?
- What risks could individuals or businesses overpromising on this technology's capabilities create for consumers and markets?

CASE FOUR: STREAMLINED ACCUMULATION AND SHARING OF PRIVATE INFORMATION

A specific example of this use case exists in the healthcare sector, where a blockchain network can be used to exchange patient data between hospitals, labs, pharmacies, and doctors and nurses. This could potentially reduce healthcare costs, improve performance, and protect patient privacy.

Discussion:

- How will this technology ensure the privacy of sensitive user information?
- What privacy policies will be put in place, and who will direct them?
- Can we rely on this technology's resilience when it is faced with a threat or breach?
- How will this benefit the social good?

Sample Questions for Further Conversation

- How will developers and the public be able to trust the reliability of blockchain technology?
- How does the traceability of blockchain transactions ensure transparency? What are the benefits of this transparency? In

what ways could it be leveraged?

- Due to the complexity of blockchains, will users find it difficult to understand and see the benefits of the technology?
- How will the security of sensitive user data be ensured?
- Will the transparency of blockchain transactions cause issues and threats to privacy?
- What risks are present to blockchain technology in terms of fraud and abuse?
- Due to the lack of regulation and mediation in blockchain transactions, how might current inequities we see in the world with wealth, social opportunity, etc. change or worsen?
- How will the high energy (and, in turn, electricity) demands of blockchains be met in a safe and responsible way?
- Will the public be able to see the value of NFTs and find good use in them?
- How might blockchains and NFTs impact the intellectual property (IP) life cycle?

As we step into increasingly virtual workspaces, a leader's humanness—her ability to listen, build meaningful emotional connections and trust—is going to be more critical than ever.

—LOVLEEN JOSHI

AUGMENTED REALITY AND VIRTUAL REALITY

When Dr. Sriji Sankar, director of emerging technology at BAB, was twelve years old, she attended space camp and still carries visceral memories of simulated zero gravity. This impression in her formative years led her down her career path. For the past decade, she's been seeking the right match to use virtual reality in BAB's training programs for medical researchers.

Her first pass was using 3-D imagery to re-create the organs of study and control animals in response to ethical complaints by the public regarding animal research. This imagery was then uploaded to a virtual reality (VR) platform that allowed researchers to study the same sample set without variance, degradation, or the deaths of more rats.

That was a win-win for the company in both steering public relations and creating a foothold in the virtual reality space. Execution, however, was a challenge. Ten years ago, headsets on the marketplace were extremely expensive. With a new program, securing funding for untested technology was a stretch, but the PR issue helped move the tech response forward.

Still, the headsets available to her were created mostly for play, so the costs crept ever higher as she built a design and engineering team to create the simulation of the animal organs. They also needed the scientists involved to get the simulations right, and that ate into the science team's research hours. BAB was trying to get a diabetes drug to the marketplace ahead of its competitors, so the political battle in the company for resources was intense.

Sriji held out, and though the going was slow, they built a strong program that not only helped with the PR issue but also became an opportunity for the marketing and sales teams. The business development specialists were able to visit physicians and administrators with their VR tools to make sales and train more industry leaders to adopt their technologies and buy BAB's equipment.

Understanding Virtual and Augmented Realities

A virtual reality (VR) experience is entirely immersive in a simulated environment. The user of a virtual reality has no interaction with the tangible reality. Think of a solo trip to the Holodeck on *Star Trek*. Everything is generated, from the surroundings to the individuals you would encounter in that reality. Because it is so immersive, virtual reality most often requires an adaptive device to enable the user to experience the simulated world.

Augmented reality (AR), on the other hand, allows users to interact with each other through a digitized intermediary—think of backgrounds in a Zoom meeting that make you look like you're in the same place as your colleagues. This is an extremely simplified example, but it is the easiest way to begin to think about the technology. A more complex example is the game Pokémon Go, in which players go to the same locations in physical reality to engage in activities in augmented reality. When a player can dispatch a lure to draw in a Pokémon for their own collection, this lure serves the entire network. In this way, the activity done in the augmented reality causes a reaction in both the augmented and physical realities by drawing virtual creatures to the augmented location and living humans to the physical location upon which the game is overlaid.

The technology for both virtual and augmented realities is complex and requires extensive visual and auditory cues and a deep understanding of human perception layered atop situational awareness. These aspects are then layered upon a software development foundation. More than legacy software development, new augmented-reality and virtual-reality world building calls on a far more robust understanding of the human experience to simulate the effects sought by the experience.

New augmented-reality and virtual-reality world building calls on a far more robust understanding of the human experience to simulate the effects sought by the experience.

THE OPPORTUNITIES

A major concern about screen time and sedentary game use in children is somewhat abated with VR interaction, in which users must get up and move to enable many of the games. Adult and pediatric congenital cardiologists in at least one US children's hospital encourage VR use in patients who need to get more moderate exercise, as many VR games engage not only the musculoskeletal system but also the nervous and endocrine systems in ways that traditional sedentary gaming cannot. Child-life specialists at the same hospital have used VR to help chemotherapy patients and other patients with semi-invasive procedures to reduce post-traumatic stress triggers by engaging the brain stem with positive interactions. This method also allows young patients to avoid excessive use of sedatives.

This physical activity is also present in augmented realities like Pokémon Go and older examples like Nintendo Wii, which react to human movement or geocaching software that uses GPS coordinates in the real world to establish credit in the augmented reality of that platform. In the case of geocaching, Pokémon Go, and other AR games, individuals and families get outdoors for fresh air and exercise.

A similar example is the gamification of fitness with tools like wireless pedometers and exercise monitors that report to a common platform on which friends and family members can compete and hold each other accountable for health goals. This augmented reality becomes deeper with graphics and images that show how far a person has walked over a set period with a geographical or altitude-based equivalent. All these augmented realities draw in human participation and can benefit users in ways that stimulate positive behavioral change.

Virtual reality offers extraordinary possibilities for learning as well. VR provided the opportunity for students to learn about biology through virtual dissection for remote coursework during the COVID-19 pandemic. VR also holds the possibility to train nuclear technicians and oil drillers to use robotics to handle dangerous energy resources without risk before they execute those tasks in an augmented reality. It's a way to manage disaster recovery training well before a disaster hits. The possibilities of VR-aided learning are limited only by our imaginations and understanding of how to create those realities in emerging software platforms.

The possibilities of VR-aided learning are limited only by our imaginations and understanding of how to create those realities in emerging software platforms.

THE CHALLENGES

For every positive aspect of augmented and virtual realities, there are counterproductive possibilities for humans. VR and AR are the successors of television, pinball games, Pong, Nintendo, and smartphone games. We already know that the more children and young adults immerse themselves in screen time, taking them away from social interactions that shape their neurological development and social skills and the critical human construct of empathy. Removing people even more from physical reality does not help with the issues we already face as a society.

The biological implications are as real as the social implications. As great as it is for people to get outside and get exercise in an augmented reality, disengagement from that physical reality can lead to a deadly imbalance. Pokémon Go sadly has a measurable death and injury count.[20] While one could make that same argument for bicycling and swimming as recreational activities that can lead to injury or death, it

20 Fiona J. McEvoy, "Six Ethical Problems for Augmented Reality," *becominghuman.ai*, December 15, 1017, https://becominghuman.ai/six-ethical-problems-for-augmented-reality-6a8dad27122.

is the nature of the injuries that happen with augmented realities that gives pause. Distracted drivers playing the game, distracted players wandering into dangerous situations, or players taking extreme risks are all root causes of violence, injury, and death in the physical intersections of augmented realities. Situational awareness while swimming or biking is a factor in conducting these activities safely. However, when we add a smartphone as an essential part of the equation for AR-based recreation, situational awareness is severely compromised by the distraction from the physical environment.

Aside from the physical dangers, financial and emotional exploitation are risks associated with any evolving technology, and AR and VR are no different. Cyberbullying has been a problem for as long as chat groups and text messaging have existed, and augmented realities are not slowing down the opportunities for this abuse. Exploitation of minors and sex workers in VR is also a criminal concern that must not be overlooked. Beyond simple ethical considerations, there are legal implications that must be addressed.

VR and AR are built upon a deep understanding of human behavior and how humans interact with realities of any type. Thus, exploitation and manipulation with things like gambling or in-game purchases tied to elevated heart rate can impact the biophysical experiences of VR and AR individually and from a psychosocial perspective.

Fortunately, we have decades of psychological and legal case studies for how humans adopt and abuse technologies to study as our realities change shape. At their roots, AR and VR are entirely built on how humans experience their worlds, and this gives us an advantage in how we want to shape the evolution of these technologies.

CASE ONE: EMPLOYEE ONBOARDING AND TRAINING

Virtual reality (VR) headsets can be used for employee training by simulating real-life work experiences. Employees can be introduced to their work environments in a safe and effective way.

Discussion:

- Will VR training compromise any effectiveness that would be present with in-person employee training?
- How will this aid equal opportunity in training? Will VR training be standardized regardless of geographic location?
- How will technical difficulties be accounted for and taken care of? Will this disrupt the effectiveness of employee training?

CASE TWO: DATA VISUALIZATION

Both augmented reality (AR) and VR can be used to enhance data visualization. Visualization can become more immersive when using 3-D and interactivity with AR and VR.

Discussion:

- Will using digital reality for data visualization introduce new bias in data interpretation?
- Is this method of data visualization user-friendly and easy to navigate?

CASE THREE: THERAPY AND SPECIALIZED TREATMENT

VR technology can be used to treat patients with specific phobias and anxiety. Stressful scenarios can be simulated in safe environments while vitals and other symptoms are monitored.

Discussion:

- Could disruptions in technology cause harm to patients?
- How will patient data be handled?
- Will HIPAA and other security regulations need to adapt to VR-backed therapy?

CASE FOUR: VIDEO GAMES AND SIMULATIONS

Thus far, the most widely encountered use cases of VR and AR have been in video games.

Discussion:

- Because VR games collect user biometric data, which is sensitive identity data, will the technology be able to protect this user information?
- How will user consent be ensured in terms of collecting biometric data?
- Will users be able to opt in or out of data collection?

Sample Questions for Further Conversation

- How can we ensure that AR and VR developers are considering the implications as seriously as the potential of their creations?
- What can we do to ensure that all children who will benefit from AR and VR learning aids have access to them to prevent deepening the socioeconomic gaps in education and access?
- What is being done to promote equity in medical AR and VR? Historically women and minorities have not been represented in studies and scientific research, so how do we bridge that gap with the emergence of AR and VR tools instead of re-creating the inequities of the current reality?
- Since VR and AR technology greatly implements AI, will the algorithms that are used produce reliable results with each new set of data?
- Will the AI behind VR and AR act predictably? Will it be precise and accurate?
- Will the collection and use of biometric data be clearly articulated to users?
- Who has access to user data collected through VR technology?
- How will VR and AR technology protect user information from data breaches?
- Are VR and AR technologies user-friendly? Are they effectively used and easily navigated by users?
- How is algorithm bias known and tested for?

- How can VR and AR be leveraged for social good?
- How much energy is needed for the development of VR and AR technology?
- What are the consequences for responsible parties when something goes wrong with the technology?
- How will emergencies and issues be handled?

When we lead, we must embrace the idea of "growth for both," that is, a holistic focus helping and enabling all employees to flourish. Flourishing employees are fully engaged and productive, propelling high levels of sustained business performance. Caring is the new competitive advantage.

—MICHAEL DULWORTH

THE METAVERSE

Dr. Da-Eun Kim is the BAB South Korean regional medical partnerships lead. She is thrilled to share the success of the doctors enrolled in their studies regarding a new therapeutic stem-cell-growth accelerator for organ tissues. She believes that the advancements the research teams have made in her region would greatly benefit researchers in Europe and North America. Dr. Kim is partnering with Dr. Sriji Sankar, the corporate director of emerging technology, to pull together technological information and merge investments to form BAB's first metaverse.

By using the same technologies Sriji has successfully used for physician training, the new metaverse team will create an environment in which the proprietary data will be part of a blockchain system and heavily encrypted. This environment will exist exclusively in the metaverse. Physicians in other regions who enroll in the metaverse experience will only be able to use the virtual lab on BAB's platform.

This proprietary control will allow all stakeholders to share information without fearing it will be stolen or misused. The metaverse allows communication with BAB hardware to set controls and enact real-life experiments using information and communication from within the metaverse. Using virtual and augmented reality models, the metaverse will allow for faster learning and hands-on simulations.

Eventually, as the BAB metaverse expands, research can intersect with different BAB and non-BAB hardware and software applications to speed up training and continuously improve global collaboration and innovation. Additionally, physicians will be able to train on real data before purchasing the equipment needed for the organ-tissue growth for themselves. Because they will train in the metaverse, once the equipment is in their laboratories, they won't have to wait for training to be able to get value from their financial investment. That expedited timeframe will make BAB the preferred marketplace leader in terms of sales options.

Understanding the Metaverse

The one thing to know about the metaverse is that there is no one metaverse. Rather, there are metaverses—plural. Eventually, metaverses may converge, but presently they are best known in the realm of gaming. In the previous chapter, I provided examples of Pokémon Go as an augmented reality, and that example could be considered a metaverse or at the edge of becoming a metaverse, depending on what defines a metaverse at any given time.

As metaverses were only imagined within the last thirty years and initially actualized in the last twelve, a clear-cut definition is elusive. It is easier to say that a metaverse is not restricted to a single technology or experience. A metaverse is not limited to or defined by a single technology but embraces as many as are both available and necessary, both existing and emerging, to enable a blended universe of the "real" and the imagined. A metaverse can expand to include AI, NFTs, cryptocurrencies, and augmented and virtual realities.

The Internet of Things (IoT) is one example of a metaverse in a home. One could use a virtual home assistant to shop; run their vacuum cleaner; manage their security system, HVAC, and Wi-Fi access; open their garage door; and start their car—all without getting out of bed. In this sense, this human being is already living in their own personal metaverse. In this example, the human does not have an avatar or digital twin; they are simply interacting in the very real metaverse of their home.

A metaverse is not limited to or defined by a single technology but embraces as many as are both available and necessary, both existing and emerging, to enable a blended universe of the "real" and the imagined.

In other instances, the human does have an avatar or digital twin in the metaverse they explore. Workers may have emerging metaverses at work. They often work with people they will never see in person, and their photographs or avatars become the widely recognized visual representation associated with them. Identity becomes a blend of video streams and static visual depictions.

With fully integrated software products, workers can access the same files and contact lists through multiple applications, platforms, and devices and share those items in real time through the cloud. The more versatile that information and data become for human interaction, the more the metaverse is expanding in those environments. An example of this is an application that is more than just email and a calendar; that application now allows immediate face-to-face calls, chat, and even group calls at the click of a button inside the software application. These calls can then be recorded, transcribed, and stored on the cloud with minimal interaction from the users. These features have become standard in most professional business applications. As the virtual walls between applications and access disappear, the metaverse emerges more fully formed.

The more versatile that information and data become for human interaction, the more the metaverse is expanding in those environments.

Social media is just one aspect of a metaverse. A metaverse offers the opportunity for richer data and more collaborative opportunities than within a single circle of friends. A metaverse takes aspects of social media and expands them to include entertainment, education, and global access to ideas and innovation.

Entertainment is the birthplace of the metaverse. From major headliners like Ariana Grande performing a concert in *Fortnite* to the mature platform Second Life, users can have meaningful entertainment experiences in the metaverse, far beyond the basics of gaming. Twitch is an entire platform dedicated to enable people to watch other people play video games—that is so meta! Yet a metaverse offers users the creative expression of world building and collaborating as they create a shared universe. From Farmville back in the early days of Facebook to Minecraft parties in middle schools, to whatever will emerge in the coming decades, the more emotionally invested users become in their virtual worlds, the stronger the hold the metaverse will have on their lives.

THE OPPORTUNITIES

A metaverse is the perfect place to simulate comprehensive learning. Imagine visiting the South Pole, the bottom of the ocean, the surface of the moon, or the deepest recesses of the living human brain. The metaverse offers unfettered access to the real and the imagined. Global collaboration at a time when we have been touched with a recent global pandemic cannot be overvalued. Climate change studies expedited through global collaboration could change the course of our current climate crises.

Moreover, cultural change might become more meaningful in a metaverse. When the information from many sources can be delivered in a data rich forum that can show, not just tell, the impacts of adverse behaviors, social movements against pollution, waste, and other inequities may have a deeper resonance for those immersed in a metaverse.

> **When the metaverse drives decisions about recording events and taking questions from virtual participants, the physical and virtual worlds have merged.**

Then there is the multidimensional aspect of a metaverse. If the avatars are expressive and the visual cues are universal, the common language of the metaverse is not limited to a single spoken tongue. Rather, the more inclusive a metaverse, the more global it will become.

Imagine if you are a nontraditional student at a major university, and you cannot go to the socials and mixers that traditional students can access. You're not in the "know" because you're working your full-time job or raising your kids, and you repeatedly miss out. Maybe you cannot attend that guest presentation that your classmates can for the extra credit your professor offered. Within a metaverse, some of those lost opportunities can be restored. When the metaverse drives decisions about recording events and taking questions from virtual participants, the physical and virtual worlds have merged.

THE CHALLENGES

With children entering adulthood whose formative years occurred in a metaverse, it is crucial that we watch these spaces and learn what we can about what is happening in them. Yet it is increasingly difficult for non-natives in the digital world to keep up with the digital natives. When children are raised in a metaverse that their parents cannot access because it defies their imagination or skills, gaps emerge quickly. Protecting children and the vulnerable in the metaverse when there is no "one" metaverse is daunting at best. This problem is no less overwhelming for law enforcement.

Bullying and stalking are complicated with the oldest technologies, and some jurisdictions still have weak laws for these crimes. When someone's digital twin becomes the victim of a violent crime, what does that mean to the physical person who has been virtually violated? To keep up, law enforcement and social services will first need to catch up with the safe, legal, and effective surveillance of existing technologies. Eventually, a plethora of metaverse interactions, some we haven't even imagined yet, will be upon us. Wherever human nature exists, crime and exploitation will coexist, and we need to accept that fact and prepare for it.

Because the concept of a metaverse is still so new and nebulous, we must assume that the challenges we face with it will be the same as any other emerging technology. We will have those who exploit it, those who are exploited by it, and the rest of us following along until it's no longer new.

CASE ONE: GAMING

In gaming, players will be able to interact with one another in a single virtual environment. What distinguishes metaverse games from VR games is that they represent a full social environment in which users can directly contact one another. Also, in-game NFT assets ensure permanent ownership of unique digital assets.

Discussion:

- Harassment, bullying, and hate speech among users are already widespread issues in video games. Will this issue intensify as gaming becomes more immersive with the metaverse?
- Could metaverse gaming have negative effects on mental health, privacy, and safety?

CASE TWO: WORKPLACE

The metaverse could very well lead to fully digital offices. With metaverse technology and VR devices, colleagues can remotely go to their offices and interact with one another.

Discussion:

- How could metaverse workplaces affect productivity, either positively or negatively?

- How will employees be held accountable for getting work done, using workplace etiquette, etc.?

CASE THREE: TRAVEL AND TOURISM

The metaverse can greatly enhance and realize virtual tourism. People will actually be present in the locations they are visiting and will be able to visit with family and friends.

Discussion:

- Can metaverse travel be used to reduce energy consumption usually produced by traditional travel (e.g., planes, cars, etc.)?
- Will users be convinced that metaverse travel and tourism are suitable replacements to physical travel?
- Will travel and tourism through the metaverse lose some of the value of physical travel?

CASE FOUR: SOCIAL MEDIA PLATFORMS

Social media will become more immersive and will evolve to almost mimic in-person interactions.

Discussion:

- With social media mimicking in-person communication, how will physical social interaction be impacted?
- As with metaverse gaming, will harassment and bullying become more rampant? How will they be regulated and monitored?

Sample Questions for Further Conversation

- Wonderland is real, and Alice has gone down the rabbit hole. How do we keep her safe from predators and exploitation? Who is policing the metaverse? Where are the metapolice?
- Encryption, identity theft, and cybersecurity are huge problems in our current technologies. As we rush headlong into new technologies, what are we doing to prevent duplicating the same problems we've yet to solve?
- Will we address social inequities in emerging metaverses, or will we exacerbate them?
- What consumer protections are in place for user privacy and to prevent nonconsensual data consumption?
- How do we vet users in a metaverse to prevent infiltration and manipulation from bots and other forms of manipulation? How do we build trust in a virtual world?
- When people are buying virtual real estate in their chosen metaverse, what will prevent scams and financial ruin?
- The more emotionally and physically invested users become in a metaverse, the more dangerous breaches of safety and sexual harassment become for those users. What do we do if users' only recourse to harassment is to leave the metaverse?

Illness needs timely intervention. Real leadership demands that we continue to take carefully calculated risks with controlled genomic and proteomic solutions to overcome current limitations harnessing the potential of technology in a collaborative environment.

—DR. ABHAY CHOPADA

GENETIC ENGINEERING

Dr. Sid Blyton, the BAB European regional medical partnerships lead, has established an agreement with researchers at two universities to find ways to improve the body's acceptance of a biliary stent using cells produced by the patient.

Because patients who receive these stents are often in ill health, it is important that the cellular matter used in the procedure be at an optimal level of health. Researchers are receiving the biomedical prototypes from BAB through Dr. Blyton and conducting experiments with stem cells to optimize the proper alignment of genetic material to improve outcomes for the placement of artificial stents in the biliary duct, supported by genetic materials, in the hopes that these therapies will lead to faster recovery rates for patients, fewer complications, and lower risks of rejection.

It's in the Genes

Genetic engineering is debatably the most controversial technology facing human beings in our lifetime. At its most essential, genetic engineering is the manipulation of a segment of DNA to influence the way in which the source cell reproduces and the traits carried forth in new cells derived from the modified genes in the new cell's DNA.[21] The manipulation could be limited to copying, or cloning, an existing cell with no additional modification. However, even a clone cell can be influenced by unintended consequences, which is what makes genetic engineering so mysterious—and potentially so dangerous.

Even a clone cell can be influenced by unintended consequences, which is what makes genetic engineering so mysterious—and potentially so dangerous.

---

21 Kevin Beck, "Genetic Modification: Definition, Types, Process, Examples," *Sciencing*, May 28, 2019, https://sciencing.com/genetic-modification-definition-types-process-examples-13718448.html.

When people first read Michael Crichton's 1990 novel *Jurassic Park*, it fit nicely in the canon of science fiction going back to H. G. Wells's *The Island of Dr. Moreau* and Mary Shelley's *Frankenstein*. Only six years after Crichton's book and three years after the blockbuster movie, Dolly the sheep was the first mammal cloned from the cell of an adult animal to enter in the annals of science, and definitive nonfiction, in 1996. Unlike the presumption of infertility made by the fictitious scientists with Crichton's dinosaurs, Dolly, the very real sheep, birthed six lambs in her life.[22]

While Dolly was the first clone of her kind, humans have been engineering genes outside the laboratory for centuries, beginning with the first domesticated dogs, cats, and livestock. The act of selective breeding gives us dogs that vary in size from a teacup poodle to a Great Dane, with noses like a pug and the long snout of an Irish Wolfhound. Even the tail nub of the corgi is genetically engineered through generations of selective breeding. Another term for selective breeding is *artificial selection*, which stands in stark contrast to *natural selection*—the term coined by Charles Darwin to explain the genetic evolution set in place by nature without intervention by humans.

As unnatural as genetic engineering may seem when we discuss cloning, it has been the inevitable progression of the science of man since the dawn of agriculture. Plants have been the target of selective breeding since the first wine was fermented, probably earlier. From roses and trees to corn and wheat, humans have been tinkering with nature since time immemorial.

The evolution of artificial selection was bound to end up in a laboratory. Long before the first sheep was cloned in a laboratory, farmers used artificial insemination to breed the most desired traits

22 The University of Edinburgh, "The Life of Dolly," *Dolly 20 Years*, retrieved December 22, 2022, https://dolly.roslin.ed.ac.uk/facts/the-life-of-dolly/index.html.

in horses and cows. No matter how postmodern genetic engineering seems, it is, at its essence, historic.

Contemporary genetic modification includes changing genes at a cellular level to control specific traits. Some stressors, such as chemicals, X-rays, or viruses, can lead to genetic mutations that may be considered favorable in certain organisms.[23] Then there is the more complex snipping and recombining of DNA to modify host cells with what is essentially man-made DNA.

THE OPPORTUNITIES

The possibility of significantly reducing cancer risk or even curing cancer through genetic engineering is not beyond the realm of possibility. By studying the genetic makeup of different groups for whom cancers are respectively more and less prevalent, gene therapies to enable the "good" genes in the populations that need them are the goal. It is better to never get the cancer than to try to stop it once it's started.

The opportunities to fine-tune our world to improve our health and chances of survival are vast.

Our bodies' ability to respond to ever newer and deadlier viruses and bacterial infections is a huge motivator for genetic engineering. Disease resistance is not just for the viruses we can catch in schools and grocery stores. We've unintentionally created antibiotic-resistant bacteria by overusing antibiotics. With an arms race between bacteria and antibiotics to fight them, the secret weapon may be in genetic engineering, both in humans and in bacteria.

Illness prevention begins with the general health of the popu-

23 Kevin Beck, "Genetic Modification: Definition, Types, Process, Examples," *Sciencing,* May 28, 2019, https://sciencing.com/genetic-modification-definition-types-process-examples-13718448.html.

lation. Being able to grow more sustainable crops with less water in changing climates is paramount to adapting to soil erosion and climate change. The opportunities to fine-tune our world to improve our health and chances of survival are vast. From saving the stem cells of newborn babies who may carry genetic predispositions to generating new stem cells for cancer treatments, we have many good uses for genetic modifications, and the potential is as limited as the afflictions faced by humankind.

THE CHALLENGES

The ethics of genetic engineering are at the heart of the debates against it. How far is too far with cloning? If cloning your dog is acceptable, what about cloning your child or spouse who has a terminal illness? When do we cross the line from restoring balance to cheating fate?

Consider the endangered black-footed ferret, whose gene pool was so limited that cloning a long-dead female was the only way to diversify it.[24] We know that humans endangered the black-footed ferret and led to its near extinction. Many species in the animal kingdom owe their demise to the excesses of human dominion.

Perhaps we bring these creatures back from the brink of extinction—or one day from extinction itself. But what about animals that never encountered humans? What about other species closely related to *homo sapiens*? Do we bring back our historic cousins? At what point does our responsibility to the animal kingdom to atone for human error end and our curiosity about the possible begin?

Yes, the arrogance of humankind led to the extinction of many species, but some died out from natural selection. In this case, the

---

24 Bob Yirka, "Black-Footed Ferret Cloned to Help Preserve Endangered Species," *Phys.org*, February 23, 2021, https://phys.org/news/2021-02-black-footed-ferret-cloned-endangered-species.html.

science fiction of *Jurassic Park* seems particularly relevant to our emerging reality.

CASE ONE: CLONING FOR THE CREATION OF A SEPARATE INDIVIDUAL

One of the most controversial uses of genetic engineering has been cloning or producing a genetically identical copy of an organism. While the ethics of cloning are hotly debated, the first-ever sheep (named Dolly) was cloned in 1996 by scientists. In 2020, scientists cloned the endangered black-footed ferret. Could human cloning be the future of genetic engineering?

Discussion:

- Could the procedure cause harm to individuals? What are the long-term effects of cloning?
- What would be the effects on diversity of genes? Could this cause inbreeding, which could in turn lead to extinction?
- What would happen if someone were unhappy with their clone? Could any next steps violate human rights?
- Could human lives be devalued if human cloning were to follow a farmlike approach?
- Would it be ethical to create a human clone that will more than likely suffer from genetically related diseases? Would this be a violation of human rights?
- Would human rights apply equally to humans and their clones?
- Would people be equally accepting of humans and their clones, or would naturally created humans be prioritized? Could this lead to discrimination against clones?

- Would a clone be able to establish its own identity? Would its only purpose be to be a replica or a replacement for someone else? Would they be expected to be like the person they were created from? Would this be a violation of human rights?

CASE TWO: CLONING FOR ONE'S OWN AESTHETIC IMPROVEMENTS

Cloning human cells can reverse the effects of aging and our aging process. The antiaging market is already a multibillion-dollar industry. Every cloned human cell is a brand-new cell. It is the precise replica of an existing cell. However, the new cell is fresh. A person could make their body younger by cloning their own unique cells and implanting those cells into the body when they are older. This method of cellular generation could greatly extend human life.

Cloning can also help improve results of reconstructive and cosmetic surgery. Without a person's own cells, cosmetic surgery may be dangerous, as medical devices are foreign objects to the human body. For example, silicone breast implants can cause immune diseases. Doctors have the skills to construct bone, fat, and connective tissue that can precisely match the patient's tissues. Therefore, if a person wants to change his or her appearance, the risks of rejection would be greatly reduced.

Discussion:

- When might the procedure cause harm to individuals? What are the long-term effects of this type of cloning?
- Could personal cellular cloning lead to unrealistic standards and expectations regarding physical appearance? What would become of diversity and self-acceptance?
- Could this further increase the disparity between social classes, given that those who are wealthy are typically those who are able to afford similar types of procedures?
- What would be the effects on society if people were to live longer lives, increasing the population?
- Could clone-based cosmetic procedures cause harm to individuals? What are the long-term effects of this type of cloning?

CASE THREE: CLUSTERED REGULARLY INTERSPACED SHORT PALINDROMIC REPEATS (CRISPR) AND GENE EDITING

The CRISPR procedure can be to adjust and manipulate not only the genes of humans but also of human embryos, allowing for the possibility of "designer babies." This could result in the "manufacture" of only certain types of people (e.g., those with a specific eye color, ethnic profile, intelligence level, overall looks and strength, and so on). While everyone wants strong, healthy babies, is using biotechnology ethical?

Livestock animals such as chickens are often engineered to grow larger breasts, which makes the animals' existence painful and almost impossible. These types of modifications make the meat better for human consumers but unquestionably add difficulty and pain to the lives of animals.

Pet breeders often attempt to use genetically limited specimens to make "purebred" lines. These animals are often riddled with health problems, largely because of the preservation of harmful genes that would have naturally fallen out of the population but persist because of preferences for certain "looks" in dogs and cats.

Sample Questions for Further Conversation

- Will the elimination of "bad genes" and preservation of only "good traits" result in people being too genetically similar? If so, what are the inevitable consequences?
- What would be the effects on diversity if gene editing became the norm? Could a reduction in diversity cause harm to society?
- Could genetic editing lead to an increased vulnerability to diseases? Could an entire population get infected by a similar kind of pathogen?
- Natural selection, evolutionary processes, and population genetics play significant roles in maintaining the biosphere. How could the earth be impacted if gene editing became standard?
- Could genetic-editing procedures cause harm to individuals? What are the long-term effects of gene editing? Could new medical issues arise?

- Would genetically modifying an embryo be a violation of that baby's human rights?
- Would this technology allow for equal accessibility? Could it further increase the disparities between social classes?
- If gene editing were to result in embryo death, would this be considered a miscarriage or malpractice?

Imagining the impact of emerging technology on our workforces and communities can be exciting and scary. As leaders, we must navigate the shifting landscape with informed, ethical choices. Our responsibility is to leverage technology to shape a more caring, human future for the generations to come.

—MARCIA MORALES-JAFFE

ROBOTICS

BAB was working on a deal to acquire delta robots to improve the speed and accuracy of their blister-pack production. These multiarmed robots use complex programming to complete innumerable multifaceted tasks. Initial investigation by BAB showed that the robots they are purchasing could measure, mold, cure, sort, fill, seal, stamp, and box ten times the quantity of pills and packs in one hour that a full production line of technicians produces in eight hours. The accuracy would make BAB the world leader in safety for drug handling, and the speed would increase their production, allowing for twenty-four-hour production. Given the difficulty of hiring enough production-line technicians, BAB was eager to put the robots in place.

However, Claudia Schlesinger, the BAB head of European manufacturing, knew from prior experience that she must consider the impact on the current workforce. She required contingency plans in place with corporate and regional human resources to address this issue before her factories could commit to installing the new robots.

While the technology is available and proven, the impacts and governance need to be addressed for the regions where BAB does business. Education programs, retraining, and opportunities to work with research teams were all pursued by Claudia and the HR Tiger Team to ensure that their workers had options when the robots begin work at the production lines. This change was inevitable, as the safety concerns and efficiencies were not debatable. The essential question for Claudia and her team was how they would handle the disruption to the labor force with minimal negative impacts.

Understanding Robotics

Robots are machines that can be programmed to operate manual tasks without direct assistance from human beings. Robotics as a field emerged from interdisciplinary efforts in manufacturing, engineering (mechanical and electrical), and computer science and has come to embrace medicine, aeronautics, and a variety of environmental applications.

Robots began in, and are still heavily used in, manufacturing, where they can perform repetitive tasks more quickly, more safely, and over time, more cost-effectively, than human beings. Robots are also used in medicine for diagnostics and surgical interventions, where their precision and speed can improve the quality and duration of life. Robots are also used to defuse bombs and enter areas of high risk to human life, where radiation or environmental exposure may be deadly to humans.

Simple examples of robotics in our daily lives are vacuum cleaners that run when scheduled, possibly when we're at work or doing errands. Additionally, self-driving cars and air-bound drones are accessible to citizens who can afford them. From gigantic robots that build and repair ships underwater to the programmable thermostat that keeps your home comfortable year round, robots are everywhere in our lives, from the bottom of the ocean to outer space.

THE OPPORTUNITIES

If you ever experienced a bad sprain or broken bone before robotic X-ray machines were available, you may remember the most painful part of treatment being the manipulation of your injury to fit into the one-size-fits-all X-ray equipment of days gone by. Even as the machines became more flexible for operators to use new angles and positions, the time for that configuration meant patients were waiting, sometimes in severe pain, both before and after the X-ray exam to ensure the pictures were clear.

Now, a robotic X-ray machine can use sensors to determine the size and proximity of the patient and rapidly capture the best image on the first try. Additionally, the results are viewable quickly by the operator to allow the patient to leave the exam and resume the next steps in their treatment. These mobile robots can drive themselves to operating rooms or intensive care units. While patients once had to be moved for X-rays, despite their precarious health, to be near the machines, the machines can now come to them. Radiologists and technicians have not lost their jobs; they can simply do them better and focus more on providing medical prognoses and medical care to humans instead of tinkering with machines. Even as machines become more precise in diagnosing tumors and injuries, they lack the ability to explain these complex situations to human patients with empathy and compassion or to explain the variety of decisions, benefits, and consequences these situations entail.

Humans have deployed robots on the space station and beneath the ocean to repair oil pipelines. They are used inside volcanoes and on the surface of Mars. Robots go places humans cannot travel safely or at all.

Another dangerous industry that is not human friendly is refrigerated meat packing. Meat cutting is an extremely dangerous job, and doing it in a suboptimal cold space for hours on end is not conducive to a healthy labor force. Likewise, harvesting and fruit and vegetable picking are labor-intensive jobs that are being replaced by robots. These jobs are typically done by migrant workers at very low wages, and the consequences of the displacement of these workers is real. However, the safety benefits and health benefits of sanitary stainless-steel robotics are the positive aspects of this complex situation.

While we are rightfully concerned with robots eliminating jobs in certain industries, robots in the classroom are poised to spark an interest in programming and engineering in the next generation of workers. Classroom robotics is done as early as kindergarten with simple directional instructions that are recorded before being transmitted to the robots. Making this technology accessible to children at the onset of and throughout their education will make careers in these fields seem tangible and less intimidating. Likewise, many school districts have access to coding software and robotics clubs, with the goals of sparking interest and providing these skills to the next generation and to encourage them to aspire to new roles in the technological future. The vivid and innocent imaginations of children paired with the power of machines are the most promising aspects of robots in our future.

THE CHALLENGES

In a 2020 mathematical modeling study, Boston University researchers Daron Acemoglu and Pascual Restrepo demonstrated that "robotics technology advanced significantly in the 1990s and 2000s, leading to a fourfold rise in the stock of (industrial) robots in the United States and western Europe between 1993 and 2007."[25] Acemoglu and Restrepo go on to explain that unlike the financial and advanced-level career advantages associated with other technologies in this book, robotics disproportionately harms the lower-skilled and midlevel workers in geographies where robotics are adopted. The loss of these jobs in lower-income geographies has a knock-on effect in the local economy that further limits the social mobility of the disadvantaged.

Unlike the financial and advanced-level career advantages associated with other technologies in this book, robotics disproportionately harms the lower-skilled and midlevel workers in geographies where robotics are adopted.

Food-service and food-packaging robots are poised to take away some of the last reliable manufacturing jobs for low-skilled workers. When this happens, where will these people work? What training is in place for them? Moreover, what will happen if those machines fail

25 Sara Brown, "A New Study Measures the Actual Impact of Robots on Jobs. It's Significant.," *MIT Sloan School of Management*, July 29, 2020, https://mitsloan.mit.edu/ideas-made-to-matter/a-new-study-measures-actual-impact-robots-jobs-its-significant.

and food production halts? All these concerns are valid as robots move beyond doing the heavy lifting of putting food on pallets and into trucks to picking the vegetables and cutting, packaging, and processing them all the way through the supply chain.

These are the implications we need to consider with skilled labor in an ever-evolving job market. Even if the job is still there, what happens when it changes so fundamentally and the demand is lower?

Let's revisit the X-ray machine example earlier in this chapter. What happens to the technicians when their work is fully optimized? Yes, patients are happier and more comfortable; they wait for less time to get appointments and spend less time at the appointments. However, at some point the critical mass of patients who experienced delays will no longer experience historical inefficiencies. At that point, there will not be a higher demand for X-rays or X-ray technicians. So how does the robot shape the future of the technician's career, and does the technician display anxiety or ambivalence about their job security when working with patients? These are the implications we need to consider with skilled labor in an ever-evolving job market. Even if the job is still there, what happens when it changes so fundamentally and the demand is lower?

Even for those who control the means of production, the current scales of robotics adoption are not balanced. Robotics are expensive investments for manufacturers.[26] This gives larger global businesses distinct advantages over smaller businesses without the capital to reap the cost benefits of robotics due to the capital expense required to make that move. This lag in the adoption of robots is seen in agriculture and meat processing where wages have never been high, so the

26 Rick DelGado, "The Negative Effects of Robots Entering the Workforce," *IT Briefcase*, October 2, 2017, https://www.itbriefcase.net/the-negative-effects-of-robots-entering-the-workforce.

investment has been a barrier for adoption.[27] Moreover, the larger the business, the more human jobs are at risk for obsolescence.

Additional concerns raised around robots are their reliance on software and the tremendous gaps we already see in security across technologies. Robots critical to infrastructure and healthcare, like any other robots using software that can be controlled remotely, are susceptible to cyberattack and therefore can be targeted as vulnerable points in society. We need better encryption and cybersecurity across all lines of technology, and robotics are no exception.

While 3-D printing is an extraordinary development and learning tool, what happens when people start making or remaking items that are someone else's legal intellectual property? When done for parody, personal use, and noncommercial reasons, it seems irrelevant, but what if it infringes on someone else's livelihood in a commercial context? When creators must combat piracy to maintain ownership of their work, we see the same patterns of social disruption we encountered with the file sharing of music in the late 1990s and early 2000s. How do we teach children in schools with 3-D printers to honor intellectual property? This is a skill robots cannot teach; robots simply follow directions.

Human beings are unique because of our ability to communicate in spoken and written language and our adaptability to changing circumstances. Currently, most robots are inflexible to change and lack the ability to adapt. For instance, if a robot-based factory is effectively producing a line of child-safety seats, putting every piece together by design, that is all good and well. However, if safety studies show that the initial design is unsafe and needs slight modifications, the robot cannot make those modifications in the way humans might recon-

---

27 Kristen Runvik, "How Robots Are Changing the Food Industry," *Food Industry Executive*, February 20, 2017, https://foodindustryexecutive.com/2017/02/how-robots-are-changing-the-food-industry/.

figure the work. Rather, the robots will require reprogramming and testing. All production will need to cease until the reprogramming is complete and the assembly can begin again. This will lead to a supply chain gap at a time when child seats will be recalled and in short supply. The inability for robots to adapt is one of the key concerns in a world that is changing rapidly. This uncertainty is a barrier for adoption when robots will become all or nothing after they take over an entire production line.

However, what happens when robots can adapt? Robots already have human speech through machine learning. As a race, humans have long pondered being replaced by the machines we make beyond the workplace. This is a question that will need to be proactively addressed to get the maximum benefits of robots for enhancing all human lives.

CASE ONE: HEALTHCARE AND SURGICAL PROCEDURES

The first fully remote surgery was completed in 2019. Before that, laparoscopic and cardiac catheterization made surgeries more tolerable. These procedures were once highly invasive and painful and carried a large risk of infection and a higher risk of death. Robots in the operating room allow surgeons to accomplish procedures with their minds and mechanical controls that their hands simply could not manage within the same time and space.

Robots can also be used in hospitals to perform more menial tasks that usually eat away at doctors' and nurses' time. This can increase the bandwidth of healthcare professionals so that they can care for more patients. Robots also can manage inventory better than humans, and that can lead to better patient safety and streamlined costs with loss prevention.

Discussion:

- In hospitals, who will be held accountable if mistakes are made by robots?
- Will patients be able to trust tasks performed by robots?
- Can the use of robotics in healthcare help prevent the spread of highly contagious and dangerous diseases?

CASE TWO: AGRICULTURE

The current agriculture industry is not equipped to meet the future demands of a rapidly increasing population. Robotics can be used to solve agricultural challenges and increase productivity. Robots can be used for tasks such as harvesting, providing weed control, mapping, fertilizing, and irrigating.[28] As robotics advance, they are able to aid in tasks that traditionally required manual skills. This change of focus in the agricultural industry should be met with further research to improve productivity and land and water conservation.

Discussion:

- How might automating agricultural tasks impact the quality of food?

---

28 Editorial Staff, "7 Major Use Cases Of Robotics In Agriculture," *Robotics*, July 17, 2019, https://roboticsbiz.com/7-major-use-cases-of-robotics-in-agriculture/#:~:text=%207%20major%20use%20cases%20of%20robotics%20in,a%20robotic%20vision%20factor%20in%20tapping...%20More%20.

- How will the role of farmers change as robots begin to complete more tasks? Will productivity improve?
- What should be done in communities to retrain and employ the vast number of agriculture workers who lose their jobs to robots on farms and in factories?

CASE THREE: SOCIAL ROBOTS

Social robots are machines engineered to interact with humans and involve social behaviors such as listening, speaking, and expressing emotions.[29] They can be used in many scenarios, including retail, education, and workplace training.[30]

Discussion:

- Can social robots be free from bias and prejudice? How could they be employed to eliminate bias from current tasks performed by humans?
- Could issues such as emotional deception, manipulation, and attachment arise from the use of social robots?
- How will the autonomy of social robots be handled?

29 Chris Wood and Susanna Dillenbeck, "What Are Social Robots? An Introduction to the Furhat Robot," *Furhat Blog*, Retrieved December 22, 2022, https://furhatrobotics.com/blog/what-are-social-robots/.

30 Furhat Robotics, "The 7 Best Use Cases for Social Robots," *Furhat Blog*, retrieved December 22, 2022, https://furhatrobotics.com/blog/the-7-best-use-cases-for-social-robots.

CASE FOUR: LAW ENFORCEMENT

Robotics can be leveraged in hazardous situations. For example, specialized robots can be used in emergencies such as navigating burning or unstable buildings and bomb disposal operations. Drones and land-based robots can be used to look for missing people in hazardous conditions and difficult terrain.

Discussion:

- Can we rely on this technology's resilience amid the unexpected challenges that can likely arise from hazardous situations?
- As robots are not as adaptable as human beings, can we trust that they will be as thorough as human beings in search and rescue where humans, and dogs who assist, use other senses and intuition to aid in searches?

Sample Questions for Further Conversation

- How will we manage our dependency on robots as they become more prolific—at what point will humans acquire a learned helplessness due to robots replacing our basic skills in life?
- Robots have and will continue to replace low-skilled human labor in manufacturing, so what is our social responsibility to the displaced workers and the communities impacted by this labor shift?
- Will robots act in a reliable way? How will their consistency be ensured in high-stakes scenarios?

- For over one hundred years, people have feared that robots might "take over the world" or "steal their jobs." How can these common worries be addressed?
- Will companies allow employees to contribute to decisions surrounding robotics and their implementation?
- For robots that interact with users, how will user data be collected and used?
- What risks do robotics pose on a physical level?
- AI is often used to enhance the capabilities of robotics, so will the algorithms that are used be free from bias and be inclusively designed?
- What kind of energy will robots require?
- How can robots be leveraged to eliminate bias and unfair treatment?
- How can the regulation of robotics increase so that accidents and risks are minimized?
- How should companies regulate robots' autonomy? Can the consequences of a robot's actions be controlled?

When leaders harness the creativity of a diverse team to dream big, enable them to be successful and celebrate their successes and failures, teams will thrive and deliver novel technologies. You never know where your next rock star idea will come from.

—RACHEL TROMBETTA

EDGE COMPUTING AND 5G

Dr. Avi Lim, BAB's South Korean regional medical partnerships lead, was approached by a national wireless provider with an appealing opportunity. The wireless development team wanted to cosponsor a BAB campus that uses their wireless 5G service, complete with an independent microwave connection and robust 5G antennae configuration to build a network in Dr. Lim's newest facility. Coupled with backup power generators, the partnership would be an opportunity for BAB to develop and test 5G-based edge computing in biotech, including radiation treatments and remote surgery, in a fail-safe environment.

As 5G was still rolling out since the 2019 launch, BAB's access was as sporadic as any other innovator who lacked their own foothold in cellular service. Using edge computing to integrate all of the monitors and pumps used in a surgery would allow the medical teams to gather real-time and predictive health information and proactively deliver medicines and interventions.

Dr. Avi invited Dr. Sriji Sankar, BAB's corporate director of emerging technology, and her team from corporate headquarters to visit with the wireless provider and discuss a proof of concept. Dr. Sankar and her team were watching the emerging field of 5G and edge computing with great interest. In their own labs, they sought low-risk opportunities to bring the promise of nearly zero latency and incredible speed to their product lines. The two main concerns they brought to the business trip were the risk of service failure and the security of the network when dealing with life-and-death situations. The teams were eager to meet and learn about the proposal, the technology, and the possibilities while weighing the risks. Hopes remain high for what 5G and edge computing can bring to the future of BAB and the customers they serve.

Understanding Edge Computing

Edge computing is a distributed computing paradigm that brings computation and data storage closer to the sources of data. This method reduces bandwidth to improve response times. Edge computing relies on faster networking to process data closer to where it's being generated, enabling processing at greater speeds and volumes. Faster communication between devices delivers action-led results in real time.

The network that currently enables edge computing is 5G. As we discuss 5G, we are specifically referencing the fifth generation of wireless networks that emerged in 2019. Now 5G technology allows computer processing to happen outside a centralized data center, whether it is a dedicated enterprise or a cloud data center. The processing of data between devices is decentralized through 5G, which holds open a continuous short-distance channel that reduces the need to funnel communication transmissions through a centralized data center. Centralized processing increases latency (delay), while decentralized processing decreases latency and improves network performance and computing efficiency.

Edge computing relies on faster networking to process data closer to where it's being generated, enabling processing at greater speeds and volumes. Faster communication between devices delivers action-led results in real time.

Today 5G is built upon a legacy of networks and is used for the wireless transmission of voice and data. The first-generation network (1G) supported the original voice-specific wireless telephones. First-generation wireless phones provided only analog voice-transmission ability and were massive devices with limited range. The 2G network followed and provided digital voice transmissions. Digital transmissions in 2G services were transferred at greater capacity than analog, providing a higher volume of calls from the enhanced infrastructure and decreasing the cost of participation for consumers.

The advent of the smartphone came with the third generation (3G) of wireless networks, in which data transmission enabled phone apps and email. With the proliferation of smart phone adoption came 4G LTE, which can handle the ever-increasing capacity required for global mobile device usage.[31] However, because the 5G network is an enhanced service

31 Qualcomm Technologies, Inc., "Making 5G NR a Reality: Leading the Technology Inventions for a Unified, More Capable 5G Air Interface," *Qualcomm.com*, December 2016, https://www.qualcomm.com/content/dam/qcomm-martech/dm-assets/documents/draft_whitepaper_-_leading_the_technology_inventions_for_a_unified_more_capable_air_interface_v.1.0.pdf.

built atop the 4G LTE network, when sparser 5G networks become unavailable, users continue to receive service through 4G LTE. This failover makes cellular data services more reliable for more users.

In the earliest days, mobile devices were simple cell phones. Then they became phones with short-message service (SMS) capabilities. The previous iterations included devices that could provide real-time updates from a variety of apps. The 5G antennae are much smaller, require less power, and are more discreet than previous cell towers, reducing or eliminating some of the service limitations of 4G LTE and earlier cell services.

In general, 5G offers greater connectivity between people and devices. It is more reliable and has fewer delays than 4G LTE. In areas where cell towers and infrastructure are sufficient, home internet service can be provided through 5G instead of, or in addition to, wired internet service providers (ISPs). With improved performance, 5G, and edge computing, devices can save directly to and back up device data to the cloud without the latency experienced with previous generations of wireless. The ability to save to the cloud means less space is required on mobile devices, so those devices can become more affordable and faster to use.

The use of 5G is rising faster than its predecessor networks, and its full potential is rapidly emerging.

The possibilities of 5G are still in their infancy, with the first iteration of the network available in 2019, and 5G supported devices followed en masse in 2020. The use of 5G is rising faster than its predecessor networks, and its full potential is rapidly emerging.[32] Presently, 5G produces less change for individuals than it does for businesses and innovators. Beyond consumer mobile devices,

32 Joe O'Halloran, "5G Set to Generate $7TN Worth of Economic Value in 2030," *Computer Weekly*, August 19, 2022, https://www.computerweekly.com/news/252523984/5G-set-to-generate-7tn-worth-of-economic-value-in-2030.

5G also provides broadband connectivity to sensors in devices from traffic lights to factory components. This interconnectivity of devices means that entire communities can function on 5G services. Additionally, 5G sensor communication promises to improve communication and reduce wait times for everything from traffic patterns to supply chains.[33]

THE OPPORTUNITIES

If a 5G sensor can detect changes in streetlight ambiance and traffic patterns, this information can be fed to the controls for the streetlights; this is an example of edge computing. No human needs to be involved; the sensors are communicating directly to the controls with no email server or other intermediary required. Responding to the data in real time means that streets can become safer for drivers and pedestrians alike. Likewise, school zones can easily be activated based on actual enrollment and disabled when children are out of school for cancellations, holidays, or changes in traffic patterns.

From factories with smart sensors to farms where cows can be monitored for symptoms of impending birth by sensors on their tails,[34] the applications of 5G are expansive and engaging. Imagine a 5G body sensor that will call emergency services if it detects a cardiac event or stroke while a person is alone and unconscious or unable to communicate. This is not out of the realm of the possible with 5G-powered edge computing and each successive generation of networking.

With 5G enabling edge computing, the Internet of Things is

33 Marco Contento, "5G and the Smart City," *Telit*, August 13, 2021, https://www.telit.com/blog/5g-smart-city/.

34 Nicola Brittain, "5G Use Cases: 31 Examples That Showcase What 5G Is Capable Of," *5G Radar*, September 9, 2021, https://www.5gradar.com/features/what-is-5g-these-use-cases-reveal-all.

more supportable than ever. Wireless connections between onboard driving systems can improve safety features and automotive performance for both cars with human drivers and cars that drive themselves.[35] Personal and home-security systems can communicate to homeowners over great distances and enable them to open doors for deliveries or pet minders. Devices not yet invented but limited only by our imaginations will eventually connect and communicate over 5G. This openness of communication between objects is at the heart of edge computing and has great potential to improve our lives.

These edge computing concepts are being tested in fully 5G-networked facilities, including a sports stadium, where devices and services are all synched with 5G to increase te speed of delivery for everything from hot dogs to ticketing and instant replays.[36] What is learned from these centers will add to the body of knowledge and the rollouts of new innovations.

This streamlining of speed with the diminishment of latency will improve the reliability of communication and allow for better production.

Inside factories, machines can communicate directly with each other rather than relying on more traditional wired networks that require relays to servers that don't provide value to the work being done by the machines.[37] This streamlining of speed with the diminishment of latency will improve the reliability of communication and allow for better production. Specifically, more attuned sensors can

35 TWI Global, "What Is 5G Technology and How Does It Work?," *TWI Global,* retrieved December 22, 2022, https://www.twi-global.com/technical-knowledge/faqs/what-is-5g#transport.

36 Samantha Murphy Kelly, "I Tried 5G. It Will Change Your Life—If You Can Find It," *CNN,* August 9, 2019, https://www.cnn.com/2019/08/09/tech/5g-review/index.html.

37 TWI Global, "What Is 5G Technology and How Does It Work?," *TWI Global,* retrieved December 22, 2022, https://www.twi-global.com/technical-knowledge/faqs/what-is-5g#smartfactories.

detect thresholds that threaten product quality and then communicate through 5G to the larger manufacturing system to delay or halt production until repairs or adjustments are made.[38] These production and supply chain adaptations are arriving at a time when supply chain struggles are entangled with record inflation. Like all things on the edge, it is right on time.

The benefits of edge computing are attainable through closer networking channels provided by 5G. That dependency on close network connections makes communication so much faster, but with that speed comes the need to process smaller amounts of data. Without a giant data center in the mix, communication needs to be simpler and less data rich in edge computing for now. Until we can bring stronger processing power to the ground level to enable more complex edge computing, the interactions will be straightforward and limited. However, with the network and the edge, limitations exist to be tested and surpassed.

THE CHALLENGES

Leading concerns around edge computing and 5G are privacy and security. As with several of the technologies discussed in this book, 5G cybersecurity is also a leading concern. When everything is on the same global network, who can hack into your baby monitor, your automobile controls, or any device holding private or sensitive information? What kind of interference can be introduced between sensors and devices? The kinds of security violations possible are limited only by the kinds of innovations developed, and we've already established that those limits are the human imagination. We need to consider

38 Ivana Kottasová, "How 5G Will Transform Manufacturing," *CNN*, March 25, 2019, https://www.cnn.com/2019/03/22/tech/5g-factory-manufacturing/index.html.

how an innovation can be exploited before it is ever sent to market if we have any hope of keeping it secure.

Privacy and identity theft are already huge concerns. The advent of 5G doesn't diminish the risk, as it brings more and more transactions and interactions to an ever-expanding pool of mobile devices. If the devices you own and use are always with you, then it is easy to know where you are based on where your devices are. This reality is already here, but it will only grow as more and more people are present on the 5G network and each future network iteration.

Equity is another concern. Retrofitting cell towers and infrastructure for 5G in rural areas that are traditionally underserved for technology is not a priority to tech companies with little to no market share in those areas. This disinterest threatens to leave underserved communities even further behind, diminishing hopes of catching up. Remote learning and the robust exchange of information in a metaverse, as discussed in previous chapters, will be ever more limited in locations lacking continuous network upgrades.

When these areas may already be hurting from job losses due to robotics and automation, a technology drought only compounds the problems of economic recession and a lack of employment opportunities. Therefore, it is crucial to consider the implications of how, when, and where 5G technologies are integrated into communities and who has access to them. Where and how edge computing and 5G can be applied to help the disadvantaged should be part of the conversation.

Additionally, 5G is compounding the issue of consumer waste, as new devices must be purchased to support the new network offerings. This moves beyond conspicuous consumption to a legitimate need to modify the technology. Unfortunately, in the race to the future, we often leave great waste in our wake. Ultimately, we must ask ourselves if we know enough about what we are doing and why we are doing it to appreciate the risks and rewards of pursuing the edge of computing and networking technologies. The benefits of speed and reliability are astounding and encouraging, but that same power is available to less trustworthy players, so cautious optimism is recommended.

CASE ONE: SMART CITIES

Cities can become "smart" by using 5G technology to connect many devices and sensors, and when managed well, the city's infrastructure can see improved transparency and efficiency. Some examples of the features of smart cities are intelligent traffic systems (ITSs), public transit, and public safety.

Discussion:

- Due to the large scale of this use case, what will be the risks posed by possible lapses and mistakes in technology?
- How will smart cities benefit social good?

CASE TWO: AUTONOMOUS VEHICLES

Technology involving 5G will accelerate the development of autonomous vehicles. It will help in terms of communication between vehicles, latency, and bandwidth.

Discussion:

- How could driverless vehicles be used to benefit the social good in terms of public health and amid the COVID-19 pandemic?

- Is it guaranteed that driverless vehicles will be safer than regular vehicles?
- Who is responsible for mistakes and accidents in autonomous vehicles?

CASE THREE: COMMUNICATION AND SMARTPHONES

The quality and capabilities of smartphones will be expanded by 5G. For example, 5G phones that have already been rolled out have better front cameras. In the future, we will see higher capacity in video-calling applications and possibly holographic calls.

Discussion:

- Will this development in communication provide benefits to society or benefit the social good?
- What section of the population will have access to this?

CASE FOUR: SMART GRID

Smart grids help enterprises to better manage their energy consumption. Edge computing will enhance these processes by providing real-time visibility into the amount of energy an enterprise is using and consuming. For example, companies can use this information to make critical decisions about their business processes in order to conserve as much energy as possible.

Discussion:

- Will the application of edge computing create more energy usage than it is assisting in conserving?

CASE FIVE: CONSUMER DATA PRIVACY

While the number of security breaches continues to rise, concern is growing within businesses that deal with highly sensitive data. Building customer trust is crucial to the success of businesses, and edge computing could be a solution. This technology provides more options regarding security and control, making it more customizable to specific business cases.

Discussion:

- Will implementing edge computing for protecting sensitive data complicate workflows more than improve them?
- Will consumers be hesitant to buy into companies that use edge computing because of their lack of knowledge regarding the technology?

CASE SIX: VIDEO SURVEILLANCE

Video is known to create the most data compared to all other data sources, which can cause high-risk security vulnerabilities due to the large amount of data being transferred over networks. Edge computing can be implemented to process this large amount of data in a timelier manner as well as keep the data more secure.

Discussion:

- Will the security benefits of edge computing and its data-processing capabilities outweigh the cost of implementing it into business processes?

Sample Questions for Further Conversation

- Considering the revolutionary nature of 5G use cases, how will the consistency of these technologies be ensured?
- Have the technical limits and potential risks of 5G-backed technologies been communicated to users?
- There have been many misconceptions and rumors surrounding 5G, such as possible health risks. As beliefs such as these become widespread, how can research be employed to curb public anxiety surrounding emerging technology?
- Will the increase of data and types of data being collected lead to privacy concerns regarding sensitive user information?
- How might 5G affect cybercrime? Will the risk of breaches and cyberattacks increase?
- What populations are 5G devices and technologies accessible to? How can they become more accessible for more areas of the population?
- How can 5G be used to develop sustainable technology that can lower energy consumption and benefit the environment?
- How will quality of life be affected by 5G technology?

- How will accidents and emergencies related to 5G technologies be resolved in a timely manner?
- Who will be responsible and held accountable for accidents, injuries, or deaths related to 5G technologies (e.g., autonomous vehicles)?
- How confident can end users be in the performance of edge computing?
- What deficiencies, if any, have been pinpointed due to the implementation of edge computing in business processes?
- Are end users made aware of the fact that their data is being handled by edge computing?
- How are the outputs of edge computing documented, considering that there is such a high level of data traffic?
- What privacy limitations exist when utilizing edge computing?
- How confident can end users be in the privacy of their data when edge computing is being used to process it?
- What new security issues may arise due to the use of edge computing?
- What types of attacks are systems more vulnerable to when edge computing has been implemented?
- How are you ensuring that new biases are not introduced due to the use of edge computing?
- What sections of the population will have access to technologies that use edge computing?
- How does implementing edge computing into business processes benefit society as a whole?

- Who is held accountable if there is a data breach in the systems that use edge computing?
- Are there processes in place to ensure those responsible for implementing edge computing are held accountable if something goes wrong?

Metrics! Metrics! Metrics! Objectively and obsessively measure where you were, where you are, and where you are going.

—TARUN RISHI

3-D PRINTING

BAB's Dr. Sriji Sankar, the corporate director of emerging technology, was asked by the chief operating officer, Sinead O'Leary, to help with a major peer-review conference. The participants came from across the globe to discuss a new artificial bile duct to improve liver function. The prototypes were well received in the national headquarters, but delays in shipping services and legal concerns about corporate espionage and lost intellectual property drove the executive team to look for new ways of sharing prototypes.

Dr. Sankar knew that 3-D printers were sent to all BAB's major laboratories within eighteen months of the conference. She collaborated with the prototype engineers and uploaded the schematics to BAB's secured metaverse, from which each laboratory could freely print their prototypes in a secure location. The prototypes then accompanied each scientist to headquarters for a detailed comparison to address any variance from the printers and materials and continuously improve BAB's new prototype peer-review process.

It All Adds Up: Understanding 3-D Printing

As you build your knowledge and technical literacy to include 3-D printing, it's helpful to know that it is also called *additive manufacturing*. The former term, "3-D printing," best describes the result—a three-dimensional object that was printed from a computer design. The latest term better explains the process, which is layer after layer being added to "additively manufacture" the 3-D object. Both terms are correct, but each explains a different aspect of the whole. Of course, the computer design informing the process must include all dimensions of the object being printed[39] to produce the expected product, and that is where the technology behind 3-D printing is most complex.

Additive manufacturing is revolutionary because the majority of large-scale manufacturing is subtractive. Subtractive manufacturing

---

39 TWI Global, "What Is 3D Printing? Technology Definition and Types," *TWI Global*, retrieved December 22, 2022, https://www.twi-global.com/technical-knowledge/faqs/what-is-3d-printing.

processes involve reductive actions—essentially cutting or stripping away through chemical, environmental, or mechanical efforts—to shape parts to assemble in a finished product or that product itself.[40]

The same computer-aided design (CAD) software that has been used in the automotive and manufacturing industries for years to create schematic "recipes" for subtractive manufacturing is the grandfather of computer-aided manufacturing (CAM) and 3-D scanners, which calculate dimensions to produce designs that can be reproduced through 3-D prints. Traditional subtractive manufacturing required for each piece and part to be manufactured separately using multiple subtractive processes (grinding, cutting, etc.) to assemble the final product as directed by the CAD design.

In the 1990s, 3-D printing began with the fused filament fabrication (FFF) process of building up filaments of an acrylic styrene material. Materials now include polymers, metal, ceramic, and glass, as the manufacturing processes now include laser technology, which makes more materials viable for the 3-D print output.[41] With the basic understanding of what additive manufacturing is and the knowledge that it produces a 3-D object, we can learn more about what it can create and the benefits of using it.

THE OPPORTUNITIES

Because of the additive process of 3-D printing, it only uses what is necessary to create final products, as opposed to the process of subtractive manufacturing. In traditional mass production, bulk items are

40 Form Labs, "Additive vs. Subtractive Manufacturing," *Form Labs Guides*, retrieved December 22, 2022, https://formlabs.com/blog/additive-manufacturing-vs-subtractive-manufacturing/.

41 Tony Hoffman, "3D Printing: What You Need to Know," *PC Magazine*, July 1, 2020, https://www.pcmag.com/news/3d-printing-what-you-need-to-know.

used, and waste is created. It is the additive nature of 3-D printing that makes it less wasteful as a manufacturing methodology.

It is the additive nature of 3-D printing that makes it less wasteful as a manufacturing methodology.

Additionally, 3-D printing is widely available to individuals. Not only do various groups share printable diagrams, but schools, clubs, and libraries across the United States also offer access to 3-D printers for free or for minimal charges. For those with expendable income, 3-D printers are becoming more and more affordable to the average consumer. Additive manufacturing on a small scale is available to the masses.

Because it creates less waste, additive manufacturing is ideal for rapid prototyping, in which a large investment in materials would be cost prohibitive. This allows the iterative development of goods with less risk of high unit cost or production failure. Within the medical device realm, this allows for faster innovation to improve patient outcomes.[42]

Unique items are also perfect candidates for additive manufacturing. Medical devices and dental implants are extremely specific to the person who needs them, and 3-D printing allows for faster delivery of highly specific items at lower costs than those required by traditional manufacturing processes. This is already producing a higher quality of life and better recovery times for patients who have personalized joint implants.[43]

The additive manufacturing of human livers, kidneys, and hearts may one day prevent the deaths of individuals waiting for organ

---

42 Form Labs, "Prototyping Medical Devices In-House with 3D Printing," *Form Labs Interviews*, September 27, 2017, https://formlabs.com/blog/medical-device-prototyping-and-testing-in-house-using-3D-printing/.

43 OrthoFeed, "Generative Design Offers Solution to Patient-Specific Knee Implants," *OrthoFeed*, October 18, 2022, https://orthofeed.com/2022/10/18/generative-design-offers-solution-to-patient-specific-knee-implants/.

transplants and reduce the risk of organ rejection by using organic material matched to each patient. These organs can already be printed in multiple materials using magnetic resonance imaging (MRI) scans of an individual's organs for educational purposes.

The applications for 3-D printing are limited only by the materials currently available to create finished objects. In the restaurant industry, viscous foods—particularly chocolate-based foods—are ideal for the type of additive layering that 3-D printing uses to create finished objects. While the current food industry is still emerging, with confectionary sculpture leading the way at trade shows, the future will include engineered plant-based meats with more meatlike textures and foods more easily consumed by the very young, very old, and those with digestive concerns.[44]

The applications for 3-D printing are limited only by the materials currently available to create finished objects.

We are at the point in time when, if you can imagine it, you can make it with a 3-D printer.[45] Beyond food, 3-D printing is also used in the fashion industry.[46] Cotton producers have produced an educational program that includes materials and fiber education along with downloadable files to print specific textures of fabric in small batches using cotton-fiber printing based on 3-D.[47] Other methods of additive

---

44 Lucas Carolo, "3D Printed Food: All You Need to Know in 2022," *All 3D Pro,* November 22, 2021, https://all3dp.com/2/3d-printed-food-3d-printing-food/.

45 Bernard Marr, "7 Amazing Real-World Examples of 3D Printing," *Bernard Marr & Company*, retrieved December 22, 2022, https://bernardmarr.com/7-amazing-real-world-examples-of-3d-printing/.

46 Mae Rice, "17 3D Printing Applications and Examples," *Built In,* retrieved December 22, 2022, https://builtin.com/hardware/3d-printing-applications-examples.

47 "Introducing Digital Cotton Fabrics: CottonWorks™," cottonworks.com (Cotton Works, July 31, 2020), https://www.cottonworks.com/en/news/introducing-digital-cotton-fabrics/?gclid=CjwKCAjwh4ObBhAzEiwAHzZYU0AaECm9FmG1sjl6amolOxcdfni3rAdaOuinYWhfop5tplb2YhpBEhoC35QQAvD_BwE.

manufacturing include fiber guns that can be sprayed on a model to make custom-fit garments directly on the body.[48] From jewelry to shoes, the revolutionary opportunities of 3-D printing are stimulating the imaginations of fashion designers at a time when supply chain issues are genuine concerns.[49]

While it once would have taken hours to get a part and a person to make repairs on a wind turbine or an aircraft, with additive manufacturing and computer files, highly specific parts can be re-created in real time at low costs to keep business moving. This innovative approach can also keep household appliances out of landfills, as we can now repair things that were once more cost-effective for us to replace. We see that 3-D printing holds great promise for a strained supply chain.[50]

THE CHALLENGES

Presently, the materials most common to 3-D printing are not as durable as those used in traditional manufacturing.[51] Durability and fragility are major concerns for materials science in the field of additive manufacturing. Another materials science concern is that we currently have thousands of material options for use in traditional manufactur-

---

48 Charlotte Hu, "Bella Hadid's Spray-On Dress Was Inspired by the Science of Silly String," *Popular Science*, July 32, 2020, https://www.popsci.com/technology/fabrican-spray-on-dress/..

49 Sculpteo, "3D Printed Clothes in 2021: What Are the Best Projects?," *Sculpteo.com*, retrieved December 22, 2022, https://www.sculpteo.com/en/3d-learning-hub/applications-of-3d-printing/3d-printed-clothes/.

50 Avalanche, "3D Printing Replacement or Replicate Part," *3dprintedparts.com*, March 31, 2021, https://www.3dprintedparts.com/2021/03/31/3d-printing-replacement-or-replicated-parts/.

51 Mallika Rangaiah, "3D Printing Technology: Advantages and Disadvantages," *Analyticsteps.com*, May 26, 2021, https://www.analyticssteps.com/blogs/3d-printing-technology-advantages-and-disadvantages.

ing and significantly fewer for additive manufacturing.[52] Of course, both concerns—durability and material variety—are at the center of future-forward research and are areas to monitor for developments.

As additive printing innovates and evolves with new materials, we must remember that we live in a world where plastic consumption and plastic waste are major ecological concerns. Presently, most 3-D printing is done with plastics, and its advancement is at odds with other efforts to reduce plastic consumption for the health of our oceans, earth, and atmosphere.[53] The emissions created by 3-D printers also require further study, as the filaments and radiation produced by additive manufacturing could pose human health risks.[54] The good news is that most 3-D printing waste (e.g., prototypes that have served their purpose or misprints) can be recycled, and the same material-science industry is investigating other ways to reduce plastic waste through plant-based components and postconsumer recycled-plastic materials.[55]

As each 3-D print is a unique item created at a particular time, the conditions of the machine, materials, and environment can impact the output. While large factories are set up to control for the highest level of quality, 3-D printing has more potential variance, which

---

52 Team Xometry, "Complete Guide to the Pros and Cons of 3D Printing," *www.xometry.com,* February 15, 2022, https://www.xometry.com/resources/3d-printing/3d-printing-pros-and-cons/.

53 Kayla Matthews, "3 Pros and Cons to 3D Printing for Medical Devices," *Greenlight Guru*, December 15, 2019, https://www.greenlight.guru/blog/pros-cons-3d-printing-medical-devices.

54 Parham Azimi et al., "Emissions of Ultrafine Particles and Volatile Organic Compounds from Commercially Available Desktop Three-Dimensional Printers with Multiple Filaments," *Environmental Science & Technology*, 2016, https://pubs.acs.org/doi/pdf/10.1021/acs.est.5b04983.

55 Printing It 3D, "Is 3D Printing Wasteful? The Facts Explained," *Printing It 3D,* retrieved December 22, 2022, https://printingit3d.com/is-3d-printing-wasteful-the-facts-explained/.

means that finished pieces may not always be identical. This is less concerning for works of art than for prototypes you wish to test or produce at a high volume.

Because additive manufacturing typically creates one item at a time and requires some type of postprocessing activity like sanding, coating, or curing, it is very fast for one item but very slow for many items. For this reason, producing a large volume of finished products may benefit from 3-D prototypes but may be most cost-effective, at this time, using traditional manufacturing methods. Software also needs to advance with production methodologies to make additive manufacturing scalable for mass production.[56] Therefore, the volume and speed of additive manufacturing output are areas to watch.

As additive manufacturing accelerates and begins to replace more traditional manufacturing at scale, we need to be mindful of the impacts on employees and communities. Printing on demand and reducing transportation impacts are huge benefits to producers, consumers, and the environment. However, with manufacturing jobs continuously shrinking along with opportunities for unskilled labor, we need to be mindful of the trade-offs when we accelerate the adoption of new technologies like additive manufacturing.

Of course, 3-D printing can be used to create less-than-desirable objects. Weapons are of legitimate concern—specifically, plastic blades or firearms that are less detectable by routine security scanning for metals.[57] Moreover, because all you need is a software design to print an object, dangerous weapon designs can be easily shared to areas

---

56 Michael Molitch-Hou, "Three Areas Holding Back the $10.6B 3D Printing Industry," *Forbes.com,* April 25, 2022, https://www.forbes.com/sites/michaelmolitch-hou/2022/04/25/three-areas-holding-back-the-106b-3d-printing-industry/?sh=6ca574bc4935.

57 Louise Gaille, "21 Biggest Pros and Cons of 3D Printing," *Vittana.org,* January 9, 2020, https://vittana.org/21-biggest-pros-and-cons-of-3d-printing.

under embargo, where traditional trade restrictions would have prohibited the export or manufacturing of arms.

Finally, copyright, patent, and trademark laws must be considered with 3-D printing. You don't want your employees to use additive printing to re-create and sell something that is going to bring litigation or loss. This is an area in which you need to ensure that you and your employees understand the implications of misusing intellectual property. You can borrow from the field of traditional manufacturing to educate yourself and navigate this concern.

CASE ONE

Currently, 3-D bioprinting remains an untested clinical paradigm and is based on the use of living cells placed into a human body.

Discussion:

- How will 3-D bioprinting (e.g., human organs and tissues) be tested?
- On whom will they be tested?
- Could the products cause harm to individuals?
- What are the long-term effects of 3-D bioprinted organs and tissues?
- Could there be new medical issues that arise?

CASE TWO

Could 3-D printing further increase the disparity between social classes, given that those who are wealthy are typically those who are able to afford similar types of technology or procedures?

Discussion:

- Could 3-D bioprinting lead to unrealistic standards and expectations regarding physical performance or enhancement?
- As this technology increases in popularity and use, could people be subject to coercion (e.g., athletes)?
- In the military context, will enhanced bioprinted organs help reduce the number of injuries and casualties caused by war through the extra protection they might provide soldiers? Or will they, rather, render wars even more lethal and destructive (for both soldiers and civilians)?

CASE THREE

Natural selection, evolutionary processes, and population genetics play significant roles in maintaining the biosphere.

Discussion:

- Can 3-D printers be used for printing mutated organs capable of providing better lifestyles?
- Would people be able to live for 150 years or longer?
- What would be the effects on society if people were to live longer lives, increasing the population?

Sample Questions for Further Conversation

- Who is responsible for weapons or dangerous materials created with 3-D printers?
- What effects will additive manufacturing and 3-D printing have on IP? Will traditional IP laws limit the adoption of 3-D printing or additive manufacturing technology by end consumers?
- How would copyright and patent infringements be monitored and addressed?

Like scientists, business leaders must look beyond the excitement of creating new technologies to the impact on individuals and society from their use. Just because you have a hammer doesn't mean you should hit something with it.

—MICHAEL FRANKEL

2013

2023
That AI filter has definitely impacted this year's intern candidate pool."

ARTIFICIAL INTELLIGENCE

BAB's chief human resources officer, Sarah Silver, was long aware of a lack of diversity in the company's technology operations. It was partly the result of a tight talent pool and academic programs that historically attracted more men than women for a variety of reasons. For traditional IT programs, the gender discrepancy did not appear to be limiting returns on investments, but for BAB's AI endeavors, it was a different matter entirely.

Dr. Sriji Sankar and Sarah Silver both attended a conference on AI use in business to better understand the technology their workforce was expected to use as the organization deployed solutions across a variety of divisions. What they learned was that AI was quite different from traditional technology. The way it was conceived, developed, and used depended heavily on the people engaged in the process of using it. If the programs were only being shaped by a narrow section of BAB's workforce, the outcomes could fall short of their potential value while also introducing risks for the larger enterprise.

When they presented her thinking to BAB's chief information officer, they learned that he had similar concerns about the diversity of people involved in the company's AI programs but for entirely different reasons. He explained how the data-science team was focused on AI's technical function, with little extra time expended on probing the risks and ethical issues that could arise after deployment. There were too few people taking part in something that ultimately involved the whole workforce.

Deliberating on the issue and consulting her peers, Sarah realized that not only did this problem have a solution but the solution could also bring about change across the organization and finally begin to move the needle on the stubborn gender disparity in BAB's technology programs. AI wasn't the problem. It was the solution.

Collaborating with the data-science teams, Sarah developed an outreach campaign to identify people throughout the business to serve on an AI advisory committee. The committee was charged with providing feedback, concerns, and opportunities to the teams developing and using the applications. Sarah also turned to a third party that offered targeted education and resources to help people build AI fluency so they could meaningfully contribute to the AI life cycle.

It did not take long for Sarah to see more and more people engaged with and excited about using AI. It was just the kind of participation that could bring more bright minds and insights to the decision-making table while simultaneously providing employees with opportunities to learn, grow, and advance the business. The very thing that she thought would perpetuate inequity became the greatest vehicle for promoting equality.

A Primer on Modern Artificial Intelligence

It is evident that artificial intelligence (AI) is the most transformative technology of the twenty-first century. It already permeates nearly every industry and business function, and its capabilities are only expanding. While AI has reached a level of maturity at which it has been adopted at scale by organizations of every type, we are still only at the beginning of our journey with these powerful tools. Where we take AI—and where it takes us—is being defined today in research labs, in the marketplace, in the halls of government, and in the day-to-day function of enterprises worldwide. To understand where we are headed and what it means for using technology, with an eye toward purpose and equity, we need to consider how this journey began and just what we mean when we speak about AI.

The vision for AI was birthed at the dawn of computer science, led most notably by famed scientist Alan Turing. The technology of the era imposed inherent limits on the types of research that could be conducted into intelligent machines. Yet by 1956, computer scientists gathered at Dartmouth University to discuss the burgeoning field and to present what is commonly known as the first AI program (called Logic Theorist). It was also at this meeting that computer scientist John McCarthy first used the term "artificial intelligence."

In the ensuing decades, computing technology advanced and permitted new and more powerful AI. In the 1980s, AI programs called "expert systems" were so mature that they were used by nearly half of the Fortune 500 companies. Yet these systems were limited in their capabilities, and while commercial adoption slowed, academic research continued.

AI is not one discrete element, such as an algorithm. Rather, AI is the result of a union of data, computational capabilities, complex algorithms, and the scientists who combine them.

The world-changing technology developments in the 1990s gave rise to hardware and software that is the foundation of today's AI: parallel processing via the CPU and the GPU, massive data sets and the rise of "big data," accessible and affordable data storage at scale, and the networking of all these things via the internet. Viewed in this way, AI is not one discrete element, such as an algorithm. Rather, AI is the result of a union of data, computational capabilities, complex algorithms, and the scientists who combine them. These practice areas, inventions, and maturing technologies opened the door to begin exploring new kinds of AI, which permit machine intelligence that can not only replicate human capabilities but, in some ways, even exceed them.

To be sure, there are many misconceptions about just what AI is, with popular narratives framing AI as something that will imminently have volition, awareness, and even some lifelike anima. But AI has no sense of self or appreciation for the world in which it operates. In truth, AI is a process of sophisticated calculations whose outcomes reflect an accurate and useful depiction of the information gathered.

Take machine learning, arguably the most exciting method of AI today. A machine-learning algorithm is supplied with data, and by its nature, it independently determines correlations and performs calculations that yield representations of functions or questions for which the AI has been optimized. Some AI models are trained to find patterns in data, others to analyze and respond to limited topic areas, and still others to decipher the contents of images or sounds.

Computer vision is a good example of how AI can think but cannot know fundamentally. When an AI model assesses an image, it does not "see" in the way that humans see. Imagine an image of a tree in a green field with a flock of birds below the clouds. To the human eye, we can understand at a glance that the birds are flying in the sky, the tree has roots in the ground, and the clouds are meaningless shapes of condensation. If the sizes of the birds are altered or the tree leaves change colors, we still know what is conveyed in the image.

AI understands none of this because its vision is based on a mathematical consumption of the image. It analyzes the image pixels. The algorithm computes the relationships between the pixels, comparing it with relationships and categories it has "learned" via model training, and the mathematical output is a high-confidence prediction that the image contains a bird, a tree, and a cloud. But it does not *know* this, nor anything else about nature and physics. The result of its calculations accurately reflects something in the real world.

This ground truth of AI applies to the many kinds of AI in use today, the following among them:

- **Computer vision.** With static images as well as real-time video, the AI analyzes visual data to determine what is being portrayed and extract knowledge from it. It can be used for things like facial recognition for security purposes, monitoring equipment in industrial settings, and fueling the numerous AI systems involved in autonomous vehicles.
- **Natural language processing.** The AI processes text or audio of common language, which includes qualities like colloquial phrasing, and it can also return outputs in a similar way. This kind of AI is used widely in the form of chatbots. Importantly, an NLP model needs to be trained with data for the setting in which it is deployed. The substance of spoken language in a clinical setting in New York is substantially different from language used on a retail store chatbot in Madrid. The underlying AI model is trained to perform accurately for a narrow application.
- **Speech recognition.** While NLP processes what is said, speech-recognition AI deciphers what the language means. This goes beyond just the literal meaning of a spoken input. It includes things like sentiment analysis, in which the AI determines whether the user is expressing frustration, excitement, or worry. Applications include things like automated call centers that field and transfer calls to the appropriate human operators or, in healthcare settings, patient intake and monitoring.

- **Planning and predicting.** Using historical and real-time data, AI can be used to plan intricate and complex processes in granular detail—and it can dynamically adjust plans in response to new information. This is useful for things like facilitating the most efficient on-demand materials availability in logistics as well as predicting when something might occur (such as a potential equipment malfunction after a certain period of usage). Forecasting the future in this way allows organizations to prepare for risks and opportunities.
- **Recommendation systems.** Internet users are accustomed to AI that predicts what will be valuable to the user and recommends it based on a variety of data, including user history, age, education, and location. This AI can cater to customer interest or demand, such as with online shopping, and it can also be used for targeted advertising.

With this basic understanding of what AI is, we can turn to the exciting potential of what it can do at scale.

THE OPPORTUNITIES

The exciting future with AI is in how people and organizations find innovative ways to apply it. We can think about the development of AI in three stages. Its early development occurred as a result of research in largely academic settings, with computer scientists probing capabilities in an effort to advance the field. The results of this research then emerged into the marketplace, where organizations engaged directly in developing and scaling the cognitive tools for specific business needs. The third stream is where we find ourselves today,

The exciting future with AI is in how people and organizations find innovative ways to apply it.

with organizations contending with not just what AI can do but also how best to use it—and to do so in a way that aligns with our expectations for the equitable, sustainable, and ethical use of technology.

In every industry, we see AI being used to develop new products or to create better products in more efficient ways. For example, whereas manufacturing high-precision parts for aircraft was once done with pencil and paper, diagrammed by hand, and developed through an iterative process of trial and error, today AI can be used to model physical objects in a virtual space, permitting rapid iterations to arrive at the most precise specifications and without the many hours and people required to achieve the same outcomes.

AI is also leading to new and better services, be they recommending products to a shopper, engaging a person across multiple communication channels, scheduling an appointment, finding driving directions, and much more. We see in life sciences, for example, that organizations are analyzing vast amounts of biomedical data to model personalized medical treatments. We also see services offered at a scale that would be nearly impossible with a purely human workforce, such as global on-demand logistics or real-time video analysis.

AI is a vehicle for new business models and the disruption and creation of entire industries while also increasing efficiency alongside capabilities. This allows the human workforce to either amplify productivity by using intelligent tools or to focus on more valuable and more rewarding tasks. The benefits are crosscutting because AI enables value for the business and consumers alike—while also giving human workers opportunities to perform satisfying work they enjoy.

The benefits of AI can uplift generations of people, and it is already one of our most valuable tools in enabling equal access to opportunity.

This is one element in AI's potential to produce positive outcomes across several domains. New medical treatments will save lives. A smarter allocation of resources will enable more people to enjoy reliable food, shelter, and energy. The proliferation of smart devices opens a world of education to people, irrespective of location or wealth. The benefits of AI can uplift generations of people, and it is already one of our most valuable tools in enabling equal access to opportunity.

The capacity to use AI to create good for all organizations and people is enormous, but for it to reach that full potential, we must contend with some complex challenges along the way.

THE CHALLENGES

Whenever a transformative technology takes hold at scale, it requires people and systems to adapt to using it to maximize its benefit and minimize its risk. The central issue is a matter of trust. Think about the electricity in houses and offices, which underlies almost every modern activity. We trust that our electrical systems are safe and reliable, and there is a general understanding of how to maintain electric systems and use them responsibly. This culture of trust around how we use electricity is not inherent. It grew out of the process of using and scaling to near-ubiquitous energy provision, and it was not without difficulties along the way.

The same process is ongoing today with AI, although we are still in the early stages of identifying the risks and learning how to manage them. To dig deeper, the notion of trust in AI can be divided into six dimensions,[58] and each element can impact potential risks for end users as well as the organizations deploying the AI tools.

58 Beena Ammanath, *Trustworthy AI: A Business Guide for Navigating Trust and Ethics in AI* (Hoboken, NJ: Wiley, 2022).

- **Fair and impartial.** Biased AI outputs are common concerns in AI applications, as unequal conclusions or recommendations can lead to unequal outcomes for end users. When AI is fair and impartial, its functions are consistent across all users and scenarios. A critical challenge is ensuring that the underlying data does not contain implicit biases from how the data was collected and curated.
- **Robust and reliable.** Small deviations in data can lead to corresponding changes in AI functions, which can lead the AI away from operating as intended. AI is robust when it can continue to function in the face of new and unexpected data that was not present in the training and testing processes. Reliable AI means that it functions consistently over time. Managing for these qualities requires monitoring and mitigating tactics after AI is deployed.
- **Respectful of privacy.** The data AI consumes often contains personal or proprietary information that, if divulged, could lead to unintended harm or legal consequences. Managing AI for privacy means considering how much confidential data is necessary for training and testing, how much needs to be collected during operation, where and how the data is stored, and whether the AI "leaks" that information when it is used.
- **Safe and secure.** Some AI applications can have real impacts on the safety of humans and assets, such as in manufacturing environments. Identifying whether and how AI could lead to harm allows an organization to mitigate risks to people and equipment, which is vital for trust in AI tools. Secure AI mitigates safety risks in a broader sense, protecting the tool, the underlying data, and its operation from malicious actors trying to cause harm or unintended issues that divulge data or lead to malfunctions.

- **Responsible and accountable.** Even as AI can function autonomously, the human operators and organizations using AI must answer for its outcomes. Trustworthy AI emerges from a structure and culture of responsibility and accountability. Each stakeholder in the AI life cycle understands their responsibilities in developing and using AI and mitigating the associated risks, and if using AI leads to unintended, negative outcomes, humans and organizations are held accountable for rectifying the harm.
- **Transparent and explainable.** One of the challenges with AI is that the process by which it arrives at an output is often opaque to the users, even to the data scientists who developed the tool. This so-called black-box problem can hinder stakeholder understanding of how the AI functions. To complicate the matter, different AI stakeholders require different levels of education. While data scientists may need deep technical understanding of AI function, business users may only need a more general appreciation for how the AI works, and executives may only need to understand its impact on their enterprises. Trustworthy AI is used in a transparent way, in which stakeholders and end users are aware of the AI, its intended function, and consent to the data it is consuming.

While not all these areas are relevant for every AI tool and use case, whether AI can be considered trustworthy hinges on a combination of these dimensions. It is up to the organization deploying AI to determine which dimensions of trust require treatment and then use that awareness to develop the processes and activities that can govern AI to its most valuable, ethical, and trustworthy potential. Managing for trust is not an ad hoc, infrequent activity. Instead, it requires the structured participation of every AI stakeholder across the organization.

Building Trustworthy AI Solutions

Looking at the complexity of AI deployment and the risks it can create, organizations need to orient their human capital, processes, and technologies around AI governance. This is important not just for capturing the desired ROI for the business but also for driving equity, sustainability, and ethics throughout the AI life cycle.

A core element is engaging a diversity of people in AI development and use. Every human stakeholder brings insights and contributions rooted in their unique lived and professional experiences. Taking an intentionally inclusive approach can help guide a trustworthy path from conceiving the use case to creating the tool and then to deploying and managing it over time. This means turning to people of a multitude of ethnicities, ages, genders, and business responsibilities. It takes more than just a data-science team to build AI that permits business value and ethical outcomes.

Engage stakeholders in a structured way, such as by setting up an AI committee charged with reviewing and recommending changes to the tool throughout its life cycle. An organization might also create new professional roles, such as a chief AI ethics officer. Yet AI programs are whole-of-enterprise activities, which is why every stakeholder requires a measure of AI fluency and understanding. Consider the kinds of education, training and upskilling, and targeted talent acquisition that can elevate an organization's workforce to steward AI to its greatest benefit.

When people, processes, and complementary technologies are aligned with a view toward AI value and risk mitigation, the organization is positioned to set goals for purpose, equity, and sustainability—and chart a detailed path to get there.

Aligning these human efforts is a matter of process. Processes for regular input and review will need to be created, as will protocols for deciding when to deploy the tool, under what circumstances it should be retired, and how future AI programs can be improved via lessons learned. As a part of this, the AI life cycle should include clear waypoints where risks are assessed and mitigated, and each step should be documented. This includes articulating the responsibilities of every stakeholder, which promotes accountability and compels people to take active roles in AI development and use.

There are also important technology investments for the AI ecosystem. While every organization contends with unique concerns, depending on AI application and deployment, there are supplementary tools and partnerships that can help validate AI outputs, explain how those outputs were reached, monitor for security threats, and guard sensitive data. Complementary technologies also include access to cloud computing (which facilitates scale), a hybrid of remote and local hardware, visualization software to help user comprehension, and a variety of services that can treat and improve discrete parts of the AI life cycle. There is no single constellation of technologies that supports trustworthy AI in every case. Rather, the organization needs to weigh priorities and risks and determine which investments are appropriate.

When people, processes, and complementary technologies are aligned with a view toward AI value and risk mitigation, the organization is positioned to set goals for purpose, equity, and sustainability—and chart a detailed path to get there. When data bias is controlled, it promotes equal treatment and access to user opportunity. When AI use is transparent and its functions are understood by stakeholders, it brings more people into the development and use of this transformative technology. And by emphasizing and structuring responsibility and accountability, the organization places the onus for ethical, trustworthy AI application on the human workforce that employs it.

Ultimately, the approaches and tactics that drive business value with AI are the same that enable business leaders to promote AI for good. Equity and sustainability in AI will not emerge accidentally. It takes intention, investment, and clear goals. By setting expectations and priorities, business leaders can set their organizations down paths with AI that are beneficial not only for the bottom line but also for the people impacted and enriched by this transformative technology.

Sample Questions for Further Conversation

- Has the organization established policies and controls to address bias and avoid discrimination?
- Who is accountable for AI outcomes? Is it more than one person, and are they all aware of their accountability?
- Which regulations and laws apply to AI for a given use case? Do these vary across regions?
- What are the criteria for retiring a tool, and who is empowered to make that decision?
- How much do end users need to understand about AI, and how are they equipped with this knowledge? Do they have a channel to provide feedback?
- Can AI systems perform accurately in suboptimal conditions?
- Is it clear how and to what extent the organization is permitted to use the data it holds?
- Are there complementary tools and protocols to address cybersecurity and data-security issues?
- Does the AI strategy reflect the business's culture and values?

Our future is so full of opportunity that it is easy to become disoriented by it. We can best navigate the emerging technical landscape by approaching it with curiosity, enthusiasm, and a constant inquiry.

—BRIJESH AMMANATH

PART THREE

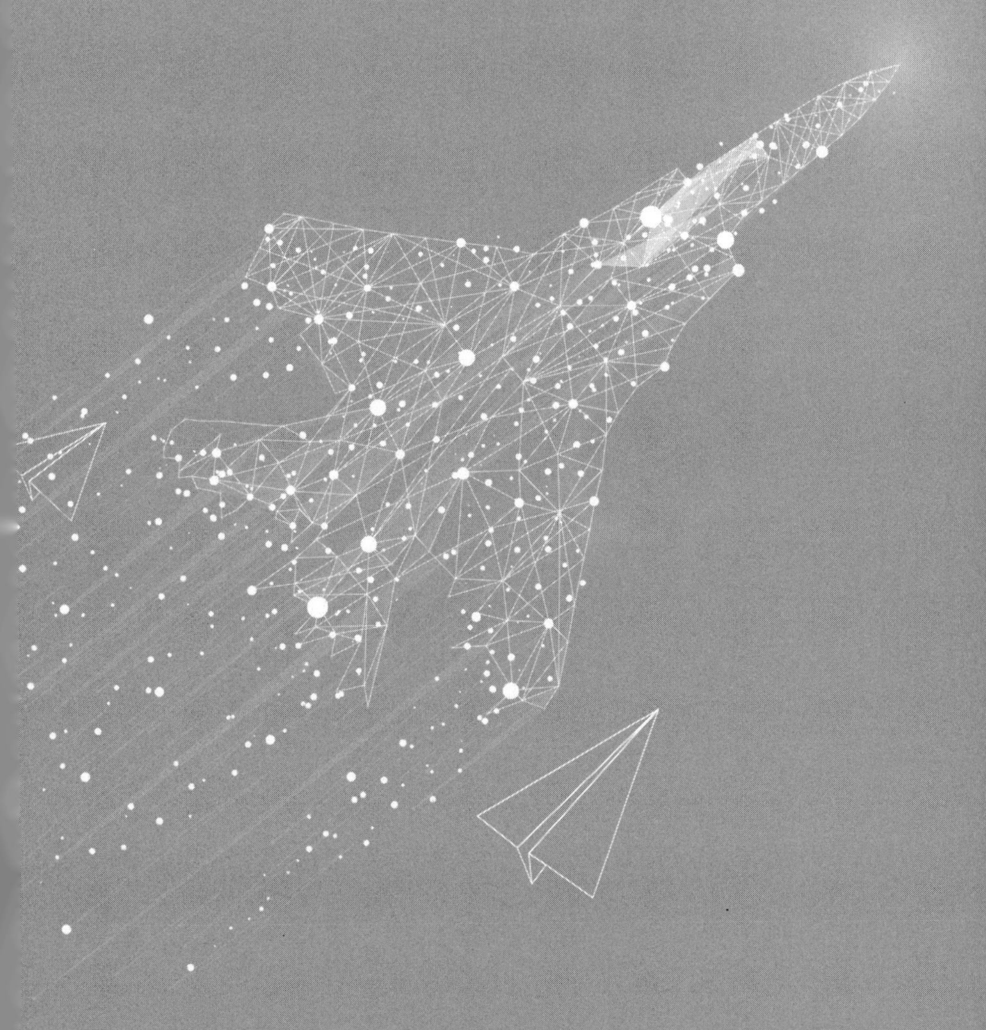

data storage is more and more necessary. Recent developments in material science have led to diamond-wafer technology that may hold enough data to support quantum computing[63] when it becomes mainstream. These innovations are driven by the needs and dependencies of others, and this connectedness is at the heart of technical literacy for a Zero Latency Leader.

We discussed robotics in part two, but robotic process automation (RPA) is the automation of tasks traditionally performed by knowledge workers. We can see the impact of this in automated software testing. Twenty years ago, software testing was a career; now it is a software package. RPA bring the questions we asked about manufacturing into the white-collar domain of the office space.

These innovations are driven by the needs and dependencies of others, and this connectedness is at the heart of technical literacy for a Zero Latency Leader.

As more and more RPA options become available, how do we realize the cost savings and efficiencies of them while considering the risks to and the implications for our workforce?

We already have cars that can drive themselves, but drivers are still required. Additionally, even the most autonomous car's cost is high, and the availability is limited. When self-driving cars are fully autonomous and easily accessible, the society disruption will be extraordinary. What happens when that is the case, and you can hire a driverless car service? What happens to the gig economy when humans are no longer required to give rides or pick up groceries? Who is at fault when there is an accident with no drivers, and how are these cars insured?

---

63 Future Timeline, "Ultra-Pure Diamond Could Store 25 Exabytes of Data," *future-timeline.net/blog,* April 29, 2022, https://www.futuretimeline.net/blog/2022/04/29-future-storage-technology-trends.htm.

Inner Space: Mapping the Human Body with Technology

Smartphones are increasingly commonplace, and smart watches and fitness trackers are not far behind. We now live longer than ever and are seeing treatments like joint replacement happening more frequently, as people are better able to survive the surgeries and recover their mobility. We already have pacemakers that can communicate over Wi-Fi today. What is next for tomorrow, when technology is no longer just wearable but embedded in the human body? This is something we need to consider as it intersects with other areas of our businesses and our lives.

The detection of BRCA1 and BRCA2 gene mutations is likely to save millions of lives through predictive early monitoring for a variety of reproductive cancers over the twenty-first century. Prenatal and newborn screenings have been shown to reduce infant mortality. These are the screening technologies we have today; imagine what we will have tomorrow. Nanorobots may be used to diagnose and treat illnesses inside the human body.[64]

Outer Space: The Final Frontier

With satellites and telescopes scheduled to orbit the sun and the moons of Jupiter,[65] we are one the precipice of a second space age. Already, private citizens are leaving Earth's atmosphere, and we are on the edge of discovering even more about the universe beyond our

64 Adrien Book, "Tech's 'Next Big Thing': 20 Technologies and Innovations to Know About from 2022 to 2030," *Honeypot*, retrieved December 22, 2022, https://cult.honeypot.io/reads/20-technologies-and-innovations-2022-2030/.

65 Future Timeline, "2030 Timeline Contents," *futuretimeline.net/blog*, retrieved December 22, 2023, https://www.futuretimeline.net/21stcentury/2030.htm.

solar system. As with the first space age, we can expect impacts to our material chemistry and manufacturing processes from what we learn from the innovations required to explore the heavens. The potential is infinite.

Beyond the Physical Plane

Part two explored augmented and virtual reality, and extended reality blends those concepts with additional accessible and commonplace devices to make the experiences more realistic for users. Extended reality drives more human beings to the metaverse to participate in games and other social encounters.

Healthcare and retail are on the cutting edge of adopting extended reality, and if you are in those industries, these are areas of particular interest. However, we will all eventually cross paths with this technology—if not in our family lives, perhaps in doctors' offices.

Because of the intersections of technology, the same concerns raised in the metaverse and AR/VR chapters apply to extended reality, but a deeper understanding of those technologies will prepare you for the emergence of extended reality.

Healthcare and retail are on the cutting edge of adopting extended reality, and if you are in those industries, these are areas of particular interest.

Beyond human representation in extended realities, digital twins are simulated replicas of reality itself. Reality in this sense might refer to a physical object like a wind turbine or building, a process like the assembly line for manufacturing a car, or a system like the financial or climate system.

The idea of a digital twin is that by creating a software-based duplicate of the real thing, a digital twin can provide an interactive way to engage with, ask questions of, or make predictions about that object, process, or system. Several ingredients are required to build a digital twin, but a few worth mentioning are 3-D interactive interfaces, live data streams, and machine-learning models for prediction.

A leading benefit of digital twins is a virtual clinical trial,[66] which enables drug development to reach people in need more quickly and at lower cost. Likewise, this methodology can be used in construction, ecology, and other types of predictive models to improve safety and reduce costs.

Past the Horizon

As we can see, there are no limits to the potential of technology. This book provides just some of the technological advances, but there are so many more. Neuromorphic hardware, bionic eyes, brain-computer interfaces, and medical nanobots are all methods of merging electronic technologies with human biology. Xenotransplantation is another method of bioengineering, using organic material from other animals in human bodies.

Hyperpersonalized technology will eventually include everything from your cell phone to an eye implant. With additive manufacturing, hyperpersonalization is an obvious next step. The sky is literally the limit.

Flying taxis are not beyond the imagination for the near future, as space tourism is already happening. Sand batteries allow for the

---

66 CB Insights, "12 Tech Trends to Watch Closely in 2022, *CBInsights.com,* retrieved December 22, 2022, https://www.cbinsights.com/reports/CB-Insights_Tech-Trends-2022.pdf?utm_campaign=marketing_tech-trends_2022-01&utm_medium=email&_hsmi=201236282&_hsenc=p2ANqtz-96Z4uZCt-bO1069A68szeZSYlv1PZS6bL5xO_uNbi91y-Z-Sz4RxgrgcTuYTYkg3GQA0m0LlaYMsUSDNgakyuxPFG6Uyw&utm_content=201236282&utm_source=hs_automation.

use of plentiful materials to store power and change the future of renewable energies.

Planning for potential technologies within your organization is crucial to taking proactive action. Above all, books like this can become your catalysts to steer the conversation in the right direction for the highest competitive advantage while driving equity, trust, and sustainability.

LEADERSHIP IN THE ERA OF RAPIDLY EMERGING TECH

Leadership requirements have evolved over the past century. Often, we equate leadership with management, but these are two very different domains that often, but not always, intersect. Exceptional managers are often detail oriented and highly organized; exceptional leaders, however, are visionaries who are highly innovative and agile.

> **Zero Latency Leaders accept the fact that the pace at which technology is coming at us is so fast that nobody can keep pace with everything that is happening—and it is only getting faster.**

We've all witnessed the optics and intensity of the modern age, and when it comes to leadership, the attributes of honesty, fairness, humility, compassion, accountability, and integrity, are nonnegotiable. We need to have self-compassion, too, so that we can pursue our passions and lead with confidence and resilience. Management skills can be learned, but the constitution of a good leader must be developed through introspection, emotional intelligence, and active listening. The journey of a good leader is to learn as much or more than we teach.

With these characteristics at the center, revisit your own skill set to integrate new skills and abilities. Zero Latency Leaders accept the fact that the pace at which technology is coming at us is so fast that nobody can keep pace with everything that is happening—and it is only getting faster. These leaders use their skill set to navigate, not manage, and thrive in the onslaught of technical opportunities.

The following focus areas will guide you. Evaluate each of these practices in your growth plan, and implement them regularly with your team to optimize your ability to meet emerging technical opportunities with confidence and mindfulness.

Become Tech Fluent

Technologists, like managers, have a domain of their own. Therefore, business leaders don't need to immerse themselves or change their focus to become technologists. Rather, leaders should focus on becoming more tech literate to be able to pursue their specific business strategy, aligned with their specific business values. Strong leaders shape an environment where technology is part of the business process and act accordingly.

Zero Latency Leaders understand not only what a specific technology does but also what its limits are and what its potential is. Then they reframe that knowledge within the decision-making chain of the company so that technology can be used to promote productivity and efficiency. To achieve this, we need leaders who are tech savvy enough to optimize their extensive knowledge on business processes to maximize technical efficiency for the company and for society. You need to use your visionary and decision-making skills to join the right technologies to natural business processes. The goal is to make work easier, not to make new work.

The technically fluent are always expanding their frame of reference to better consume and understand new information. When leaders become tech literate, they will be able to understand how, where, and why to use tech to drive more efficient decision-making.

Collaborate with Your Competition

We began this chapter affirming that the rapid pace of technological change is beyond any one person to keep up. It's also beyond the capabilities of any one company or entity to do the same. Collaborating with your competition, learning from each other, and sharing best practices are the only ways you can move forward as an industry and keep pace. Trying to do it all yourself in the modern technical landscape is no longer optimal on the pathway to success.

The pace of technological advancement is rapid. The definition of traditional competitor is constantly being challenged. To keep up with the different use cases of the technology within any industry while also addressing the unknown potential risks is a Herculean task for any single entity. Active collaboration is the only way to thrive, which often includes collaborating with your competitors. None of these technologies will come with the necessary regulations in place. By collaborating across your industry, you can inform and shape regulations that will help all entities to thrive.

Build Diverse Teams

As the world is ever more connected and transparent, for a business to thrive, it must listen to its customers and its critics. Diversity is intersectional; it is not simply checking a box and moving on. Rather, diversity involves valuing the perspectives, life experiences, cultural norms, and wisdom that a multifaceted workforce can offer. When everyone thinks alike, opportunities for novel approaches are lost. The more diverse your workforce, the stronger the engine of innovation and growth will become.

People who grew up in different regions under different circumstances will bring different problem-solving skills and approaches to innovation. They see things that are not obvious to others, and this stimulates conversation and improves your products and solutions. Because diverse employees bring new ways of thinking to business, they not only help leaders avoid common mistakes but also can tap into entirely new markets and build brand loyalty. When using technology, it is important to understand how the product will be used by the consumer. Without diversity in the organization, you will miss the mark.

Because diverse employees bring new ways of thinking to business, they not only help leaders avoid common mistakes but also can tap into entirely new markets and build brand loyalty.

Cultivate an Ecosystem

Trying to understand and keep up with the full onslaught of technological innovations and changes is akin to boiling the ocean. We have to take it one step at a time, and we need to work together. No one can do it alone.

Like the BAB example in the 5G chapter, collaboration across domains with different businesses can accelerate the promise of innovations. Ecosystems between universities and private companies or government are established examples, but your ecosystem may also include your competitors, research groups, industry alliances, service providers, start-ups, and venture capitalists.

Building a technology ecosystem across businesses is the modern equivalent of the stone-soup folktale in which everyone provides one or two essential ingredients and the whole is greater than the sum of its parts. A strong ecosystem begins with a human-centric approach to ensure the technology brought into the mix serves a purpose to those who will consume it and is not just the next "big thing."[67]

Focus on Humans with Technology

What if, instead of the fear of replacing humans with technology, we embrace humans working *with* technology? What special skills and knowledge does your workforce bring that can be augmented with technology as opposed to being replaced by it? When considering new technologies, think about how it would be *better* with input and assessment from skilled professionals.

In the most tangible example of this, biomedical devices can work with the human body to restore movement and abilities lost to injury, illness, or birth defects.[68] Police officers can use surveillance equipment and artificial intelligence to improve crime prevention and be in the right place at the right time. Begin your own practice of framing the idea of "human *and* machine" rather than "human *or* machine" to find the most effective ways to integrate technology into your business and establish a new working model that balances the value and efforts of humans with the benefits of technology.

---

67 EY Americas, "Four Essential Ingredients of Successful Ecosystem Partnerships," *EY.com,* February 24, 2020, https://www.ey.com/en_us/consulting/four-essential-ingredients-of-successful-ecosystem-partnerships.

68 EY Americas, "Four Essential Ingredients of Successful Ecosystem Partnerships," *EY.com,* February 24, 2020, https://www.ey.com/en_us/consulting/four-essential-ingredients-of-successful-ecosystem-partnerships.

Learn to Fail

When we accept the premise that we can never know everything, much less master everything, in our current technical landscape, we get comfortable with *learning* over knowing as part of our leadership style. When working to scale, failure is simply the lesson that leads to the next success.

Embrace the fact that failure is the revelation of the unknown, and once it is known, you are empowered to move forward. By normalizing failure in your business enterprise, people will stop seeking perfection and begin iterating and collaborating to accelerate success. Failure is part of the scientific process, and being proven wrong about a hypothesis or an assumption is not a moral failing. Rather, the failure of assumptions is the surest path to truth and knowledge.

Iterating and progressing with smaller stakes allows you to fail on a smaller scale. If you "bet the farm," you can lose the farm. Rather, if you risk one egg or a dozen eggs at a time, winning those wagers will become more consistent and build confidence. The old colloquial wisdom of "Don't put all of your eggs in one basket" is evergreen in business. Diversifying your approaches to risk, loss, and failure empowers you to act boldly on many fronts and then follow the most promising path forward.

Use Technology in Your Personal Life

How can you find practical uses for technology for your employees and your customer base if you don't use technology yourself? Using new and emerging technologies at home and in your personal life will help you build the context that increases your technical literacy and improves your aptitude for new knowledge. Watch and study the people in your home and community. How are they using some of the new technologies, and how does their use of technology benefit them and the community?

When you observe kids or young adults using technology, what does that tell you about applying it to your business? When you use technology routinely and make it part of your everyday thinking, it is easier for you to imagine humans with technology in the workplace and the customer space. Leveraging technology in your personal life will stimulate your imagination, your aptitude, and your ultimate leadership success.

Leveraging technology in your personal life will stimulate your imagination, your aptitude, and your ultimate leadership success.

Don't Make It a Zero-Sum Game

A Zero Latency Leader isn't solely focused on the bottom line of their business. Rather, they have a distinct awareness and affinity for the communities in which they operate. They understand the impacts of their operations on their employees and the people beyond their places of business. Responsible leadership means seeking to improve and serve the communities that are home to your business and to never exploit them.

Do you have multiple facilities? Have you visited all of them? Do you know the crime statistics, the school success rates, the transportation challenges, etc. that impact the places where your employees and customers live and work? Just like a Zero Latency Leader has the courage to question their soft skills to improve them, the same leader will look at what they don't know about the communities in which they operate and improve upon this knowledge so they can act in ways that benefit others.

For example, can you share security cameras with your small business neighbors who can't afford them for their own safety and the safety of the community? Can you offer free access software you've made to nonprofits that cannot afford the same tools? Can you share space in your infinite data-storage capacity with students who can't afford the same? What can you bring back to the community as you innovate for your competitive advantage?

Zero Latency Leaders bring communities and society along with them by sharing their learnings and innovations through scholarship, investment, and philanthropy so that they don't leave communities behind or sacrifice the future for quick wins.

Continuously Upskill Your Team

Invest in your team. The same skills of empathy, decision-making, emotional intelligence, and critical thinking you already possess should be pursued by everyone in your organization. Not everyone needs to aspire to a similar leadership role, but everyone needs to share and act within the same context of deliberation and learn to move with the speed and agility required to navigate the rapidly changing technological landscape.

Whatever your learning journey embraces, make sure you bring your team along with you. They will broaden the ideas and opportunities available to your organization through their own learning and unique thinking. Build a stronger organization by taking the Zero Latency Journey together.

Upgrade Your Soft Skills

The essential abilities to build relationships and establish trust are dependent first and foremost on your soft skills. Technology skills come and go and are limited resources. However, unlike a certificate to administer an application that no longer exists, soft skills never expire. Negotiation, communication, time management, and creative thinking skills never expire, though without practice, soft skills can get rusty.

Make a concerted effort to enhance your soft skills through practice, training, and evaluation. Classes are available, mentoring is accessible, and opportunities abound to improve soft skills across any organization. Identify where your soft skills could benefit by asking the people you work with the most and those you work with the least, and see what gaps you find and where you are the most consistent. The courage required to confront and grow is exactly the type of quality that empowers a "Learn to fail" attitude and makes a Zero Latency Leader different from one who will not look more deeply at all systems, including our own ability to interact with other human beings.

Be a Responsible Citizen

Technology can scare some people and enthrall others. The possibilities are endless, but so is our responsibility for the decisions we make. The famous quote from *Jurassic Park* springs to mind: "Your scientists were so preoccupied with whether or not they could, they didn't stop to think if they should." Always ask yourself *why* before you ask anyone else *how* to use a technology. Be mindful, be compassionate, and be accountable. Keep sustainability, trust, and equity at the center of your decision-making process.

That is what it means to be a responsible citizen—balancing opportunity and consequence to the best of our ability.

We need to find a happy medium where we advance our business, society, and humanity without compromising our future or our integrity. That is what it means to be a responsible citizen—balancing opportunity and consequence to the best of our ability. This is at the heart of a Zero Latency Leader—we choose wisely, and we accept responsibility for our decisions.

CONCLUSION

We have reviewed the historical and academic contexts of how society adopts and normalizes innovation as well as the foundational leadership skills you need now and in the future with evolving technology. We learned about the concepts of equity, sustainability, and trust in a corporate and technological capacity. The resources for these historic, scientific, and social issues are available to everyone through reading and academic study, and ignorance is no excuse for repeating the mistakes of those who came before us.

Then we discussed the different types of pressing innovations to establish a framework of technical literacy. Along that path, we sparked ideas on how some of these emerging technologies will impact your business today and in the future. We also asked pressing questions about how these newer innovations fit into our broader social and leadership constructs. We challenged ourselves to assess both the opportunities and the challenges so we can make informed decisions with confidence.

In this last section, we previewed additional technologies for you to watch and bring into your own personal frame of reference. Finally, we revisited and expanded on the leadership attributes you need to make strategic and informed choices about emerging technologies.

Now What?

Leaders have an intrinsic bias toward action. The first step is to learn about emerging technologies and leadership skills, but it is paramount to put these into practice. In the previous chapter, we highlighted focus areas on which to concentrate and build aptitude, and in this chapter, we introduce some practical and specific actions you can take now.

- Ensure that equity, sustainability, and trust are all cornerstones for your company's mission. It they are not, partner with groups that can provide insight and tools to improve this mindset and open the way for a more receptive environment. Do not become your own biggest obstacle by ignoring this vital dimension of your business.
- Do not fear the future; plan for it. Make sure you have smart goals for yourself, your team, and your business to move the needle on your technical aptitude and readiness to respond with Zero Latency Leadership.
- Subscribe to a technology newsletter or magazine that will enhance your technical aptitude. Consult some of the citations in this book and in other informative pieces you read for solid resources.
- Take a class in any of the leadership areas you need to build. Then put what you learn into action immediately.
- Challenge your team to do the same as you, and schedule regular learning sessions to share and discuss how emerging technologies can benefit your business, your employees, and your customers.

- Invest where it matters the most: fund training, access to journals and other learning aids, and participation in conferences and forums that will expand your organization's trust culture and technical aptitude.
- Use this book and the questions at the end of the technology chapters to quiz yourself and your teams about your collective understanding of the impacts and benefits of emerging technologies.
- Constantly assess how ready your company and your staff are to meet innovation and bring it into your company in a sustainable, equitable, and trust-driven manner. Then take action to improve. Repeat.
- Hold yourself to a higher standard. Remember, Zero Latency Leaders are not here to merely make quick decisions for the best business gains. We are here to make a difference in the trajectory and impact of how technology is used in business and beyond. The decisions you make today and the consequences or rewards therein are your essential legacy as a leader and as a human being.

In closing, we owe a debt to every inventor, innovator, scientist, and thinker who came before us to learn not just from their brilliance but also from their failings. We owe those who will follow in our footsteps the presence of mind to remove and avoid putting obstacles in their path. We also owe our progeny the foresight and wisdom to set an example of how one can move with agility without creating chaos, if we progress while mindful of the fairness, longevity, and resonance of our choices. Everything is a choice. Let ours be choices that make meaningful and lasting differences.

ABOUT THE AUTHOR

Beena has been at the forefront of emerging technologies for some of the largest and most prestigious enterprises around the world. A computer scientist by education, she has held technical and leadership roles with companies such as Deloitte, Hewlett Packard Enterprise, General Electric, Thomson Reuters, Bank of America, British Telecom, Honeywell, and E*TRADE. She is the founder of Humans for AI, a nonprofit focused on increasing diversity and inclusion in AI. She has also worked with start-ups, taught as a professor, supported a variety of nonprofits, and served on several executive boards. In all these roles, she has helped organizations pursue tech transformations that deliver innovation, competitiveness in the marketplace, and the equitable and ethical use of emerging technology.

Beena is a sought-after executive who asks the right strategic questions, foresees risks and opportunities, provides expert guidance, and builds equitable tech products and solutions. Alongside her professional leadership, she is a prolific writer, public speaker, and thought leader. Her research and analyses probe the ethics of technology applications as well as the necessity for diversity and inclusion in the evolution of tech. In her prior book, *Trustworthy AI*, Beena explored the characteristics of trust and ethics in AI and presented the considerations and best practices executives must weigh when using the power of AI.

With such a storied and diverse career, it is no surprise that Beena has received honors and awards for her contributions to technology and philanthropy. She was named UC Berkeley's 2018 Woman of the Year in Business Analytics, named *San Francisco Business Times*'s 2017 Most Influential Women in Bay Area Business, featured in WITI's Women in Technology Hall of Fame, featured in the National Diversity Council's Top 50 Multicultural Leaders in Tech, named CIO.com's and Drexel University's Analytics 50 Innovator, named one of Forbes's Top 8 Female Analytics Experts, and awarded the World Women Leadership Congress's Women Super Achiever Award.

Beena has the uncanny ability to see how elements of technology and community intersect to become purposeful opportunities. She thrives on envisioning and architecting how data, and technology in general, can make our world a better, easier place to live for all humans.

Connect with Beena at https://beenammanath.com/.

ACKNOWLEDGMENTS

A book doesn't write itself. It begins with a seed, a little germ of an idea that grows until it needs to be transplanted from the writer's brain to a container where it can be shared with the wider world. However, to get to that point, many phenomenal people are involved in the process.

First and foremost, I want to thank my mentors, Kwasi Mitchell and Dave Couture, for engaging in the initial conversations that led me down this path. Without their encouragement and unwavering support, the ideas that eventually became this book would not have taken flight.

Many thanks to my colleagues Alicia Rose, Joe Ucozoglu, Jason Girzadas, Dan Helfrich, Matt David, Jon Raphael, and Amy Shaw-Feirn who demonstrate and inspire zero latency leadership every single day. Thanks to Mike Bechtel, Will Bible, Debbie Rheeder, Jon Weber, Sean Page, Sachin Kulkarni, Adnan Amjad, Kate Schmidt—without their insight, this book wouldn't be what it became. I would also like to thank Anuleka Ellan Saroja, who generously provided research assistance and helped track down key sources. Her contributions were essential to the success of this project.

In a career spanning nearly three decades, I have had the fortune to learn from numerous leaders and see Zero Latency Leadership in action, across the business, nonprofit, and academic worlds. Thanks to Jude Schramm, John Wang, Vince Campisi, Tony Thomas, Meg Whitman, Karl Mehta, Keith Sonderling, Ana Pinczuk, Brenda Wilkerson, Amy Fleischer, Debjani Ghosh, Jamie Miller, Silvina Moschini, Zenia Tata, and Colin Parris.

Thank you to my friends for being the shining lights in my life and for helping me to become the best version of myself: Jimna and Cisto Cyriac, Dianne and Danny Lim, Aparna and Aditya Bhandarkar, Pooja Walke, Mary John, Sunil Palamuttam, Sameer Sayed, Uday Shinde, Rajesh Subramanian, Jojen Kochery, Zia Virmani, Neha Damle, Hari Menon, and Prajesh Kumar. I am so lucky to have you all as my friends, and I am forever thankful for your love and support.

I would also like to thank my entire team at Forbes Books who helped me shape, refine, and fine tune the ideas in this book: Lindsey McCoy, Nate Best, Amy Saad, Steve Elizalde, Evan Schnittman, Analisa Smith, David Taylor, and Heath Ellison.

I owe my life's work to Nikhil, Neil, and Sean, the most important people in my life, for always giving me the support and encouragement to pursue my passions and excel.

Finally, I want to thank all the readers who have supported me and my work. Your enthusiasm and passion for my work has meant the world to me. Thank you all for making this journey possible.